AN ILLUSTRATED GUIDE TO THE

TECHNIQUES AND EQUIPMENT OF

ELECTRONIC WARFARE

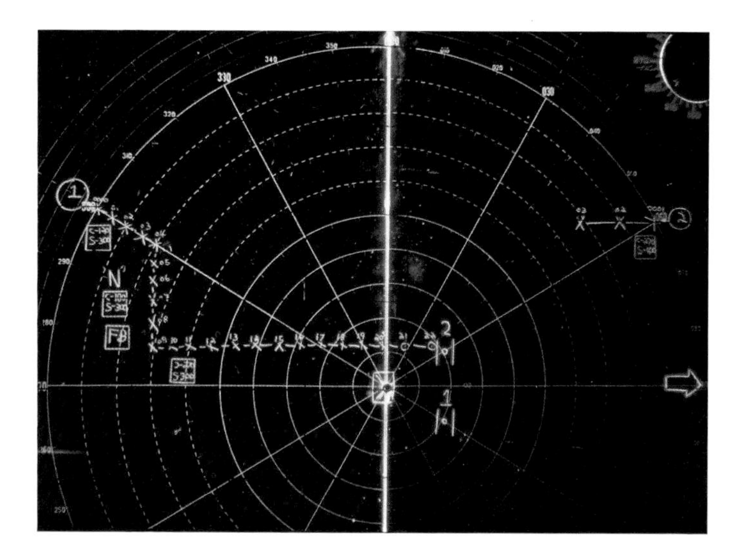

Published by Arco Publishing, Inc.
NEW YORK

611

AN ILLUSTRATED GUIDE TO THE TECHNIQUES AND EQUIPMENT OF ELECTRONIC WARFARE

Doug Richardson

A Salamander Book

Published by Arco Publishing Inc.,
215 Park Avenue South,
New York, NY 10003,
United States of America

Author: Doug Richardson is a defence journalist specializing in the fields of aviation, guided missiles and electronics.
Editor: Bernard Fitzsimons
Designer: Michael Osborn
Photographs: The publishers wish to thank all the companies and other organizations who supplied illustrations used in this book.
Typeset: The Old Mill, London
Printed: in Belgium by Proost International Book Production, Turnhout

Library of Congress Cataloging in Publication Data
Richardson, Douglas.
An illustrated guide to the techniques and equipment of electronic warfare.
(Illustrated military guides)
"A Salamander book."
1. Electronics in military engineering.
I. Title. II. Series.
UG485.R53 1985 623'.043 85-6017
ISBN 0-668-06497-8

Contents

Introduction 6
The Electronic Battlefield 8
The Electromagnetic Spectrum 12
Radar 16
Millimetre Waves 26
Infra-red 28
Electro-optical Systems 32
Sonar 34
Stealth Technology 40
Electronic Intelligence Gathering 50
Intelligence-gathering in Space 64
Countermeasures 68
Warning Receivers and ESM Systems 70
EW 'Expendables': Chaff, Flares, Smoke and Decoys 84
Active Jamming 96
Deception Jamming 100
IR, EO and Sonar Jamming 112
Communications Jamming 116
Anti-radar Weapons and Aircraft 120
Electromagnetic Pulses 140
Unconventional Threats 144
Glossary 150

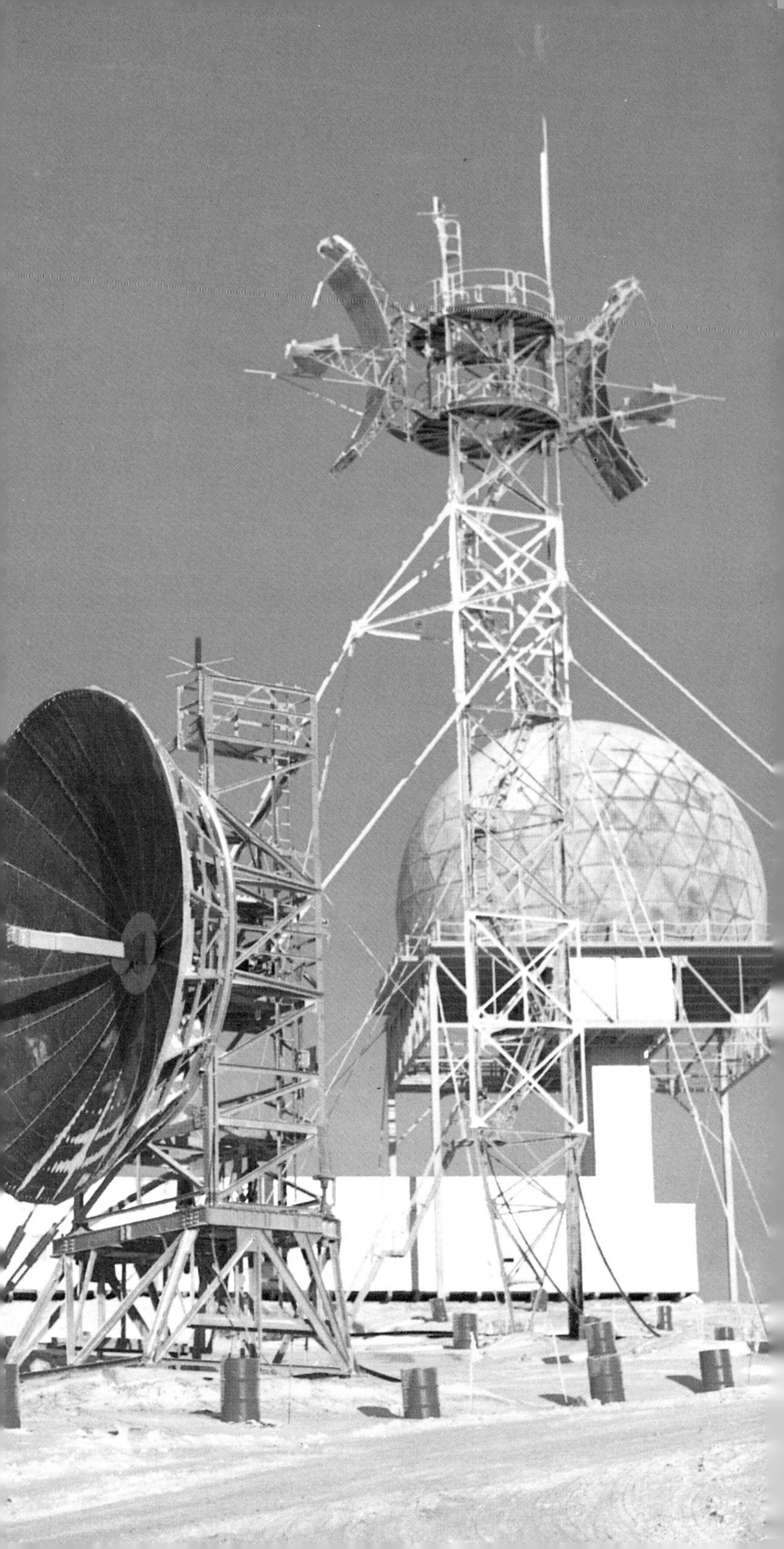

Introduction

FEW FORMS of military activity are more closely guarded than the world of electronic warfare. As far as most companies and users are concerned, the less there appears in print on this subject, the better. Like the worlds of espionage, nuclear weaponry and elite forces, "black-box warfare" rarely escapes from an all-enveloping shroud of secrecy.

At most defence exhibitions, company public relations staff are only too glad to talk to journalists and to extol the virtues of their firms' products, but this is not the case with EW. Researching the present book, the author was given by one US company a "technical brochure" whose contents were minimal – several photographs of avionics units and a boldly-printed announcement that the information in the brochure had been cleared by all the relevant military and security authorities. The fact that much more information had been published in a recent edition of a well-known US aviation magazine was dismissed with a shrug. On another stand, the mere display of interest in a photograph of an EW system was enough to provoke a brusque interrogation by company staff. Requests for information already openly published in unclassified literature led on

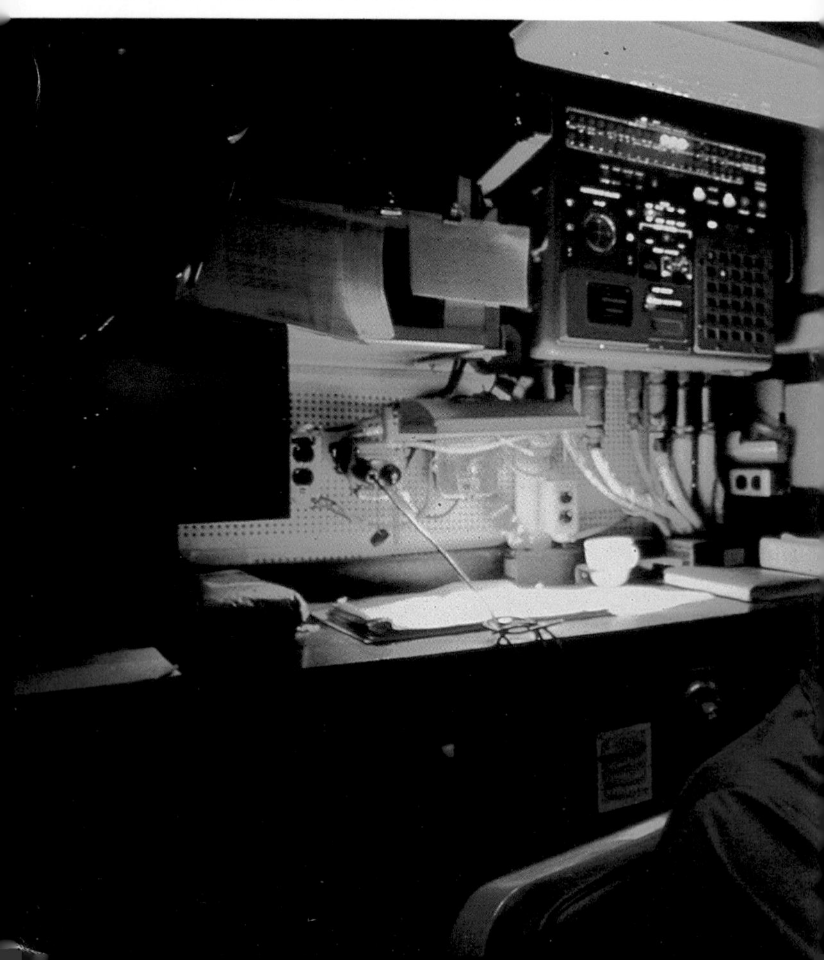

several occasions to statements that "security" would have to be consulted.

The result of such attitudes is often a mind-numbing lack of efficiency, with secrecy being equated with operational effectiveness. More than a decade after the Israeli Air Force paid a high price in human lives, Phantoms and Skyhawks for trying to counter Soviet continuous-wave threats with US-supplied countermeasures equipment designed to cope with simpler pulse-radar threats, secrecy still stifles ECM discussion. Days before these words were written, the public relations office of one major NATO air force expressed surprise that jamming pods capable of dealing with infra-red threats even existed.

The contents of this small book can serve as little more than a primer to the world of ECM. Gleaned entirely from unclassified sources such as defence exhibitions, defence journals, and the few textbooks and magazines devoted to electronic warfare, it will contain no surprises for the world's intelligence-gathering services, but should give the reader some idea of the role played by the pods, antennas and black boxes used in this most arcane of electronic arts.

The Electronic Battlefield

"THIS IS the way the world ends ...," wrote T.S. Eliot in *The Hollow Men,* "...not with a bang but a whimper." For the Royal Navy's Type 42 destroyer HMS *Sheffield,* the first sign that its world was about to end was not a bang but a chirrup — a faint whisper of sound in an electronic warfare operator's headphones. *Sheffield* now lies at the bottom of the South Atlantic, a casualty of a direct hit from an Exocet surface-skimming missile during the 1982 Falklands War, and the secrets of its last battle sank with it or are buried deep in Admiralty files in London's Whitehall.

But not all its secrets are covered by the ocean or the "Secret" stamp. As a demonstration of the capabilities of its TASS electronic warfare simulator, Canadian Astronautics Limited used the equipment to recreate the last moments before *Sheffield* was struck.

The complex systems used in electronic warfare (EW) to listen for hostile radar signals rely on computers and high-speed data processing for the analysis and identification of the resulting signals, yet valuable clues can sometimes be gleaned by simply listening to the pattern of the signal pulses — to an experienced EW operator the sounds are in many cases as distinctive as the voices of his colleagues, and equally easy to identify.

If the EW operators were listening on the afternoon of May 4, 1982, the radars of the British Task Force would have provided a steady background of sounds — bursts of squawks and buzzes as the scan pattern of each radar briefly illuminated their ship. The first clue that *Sheffield* was in trouble was a faint but regular chirrup caused by the Agave radars in the two Argentinian Super Etendard fighters searching for targets. The Argentinian pilots quickly selected the largest target

Above: Radar operators aboard the aircraft carrier *Invincible* carry out a surface search using an equipment console fitted with a circular PPI (plan-position indicator) display.

Below: Combat information centre of the US Navy's Knox class frigate *Badger*. If denied the effective use of radar, sonar and radio by enemy EW, the ship would be unable to fight.

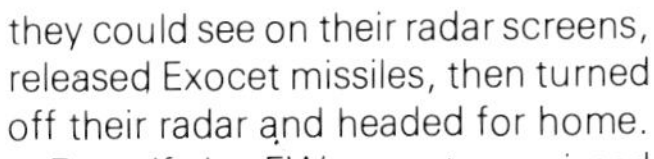

they could see on their radar screens, released Exocet missiles, then turned off their radar and headed for home.

Even if the EW operators missed this fleeting warning, their equipment should have noted a new signal and brought it to their attention. The electronic "fingerprint" of the Super Etendard's radar should have registered as a hostile emitter, sending the ship to battle stations. *What went wrong?*

The answer to that question is still classified. One account leaked to the Press suggests that the signal from the French radar was sufficiently close to that of the fleet's own Sea Harrier fighters that it was mis-identified. Another claims that someone forgot to modify the EW system's programming to delete the radar characteristics of the French-built Super Etendard and Exocet from the built-in list of "friendly" weapons and add them to the threat list.

Exocet makes its approach under autopilot control, turning on its radar seeker head only in the final stages of flight. The next warning of impending disaster was an intermittent coarse whistling sound — the signals from the Exocet seeker as it scanned for targets. Within a second or so, this would have changed from intermittent to steady, a sign that the seeker had locked on — only a handful of seconds before impact — and that *Sheffield*'s operational life was effectively over.

The birth of electronic warfare

The Royal Navy's destroyer fell victim to a missile, but its loss was indirectly due to what Winston Churchill called "Wizard War", the electronic battle fought by the technicians and engineers of both sides. World War II was the first to see the large-scale use of electronic technology and the birth of effective electronic warfare systems. Four decades later, electronic systems and electronic warfare are important components of land, sea and air combat.

Without electronic aids, the modern type of highly mechanized and automated warfare would be near impossible. In this new form of warfare, the weapons may not be the familiar tanks, ships and warplanes, but the penalties for error are as high

The electronic battlefield

A panorama of an EW battlefield based on a Middle East setting using information supplied by IAI Elta Industries Ltd. The advancing force on the right has an HQ (1) and numerous radio links and electronic sensors. The links include: from the main HQ to subsidiary HQs (2), infantry to tank (3), recce HQ (4) to foot patrols (5), and artillery observation post to gun lines (6). Electronic sensors include radars for air defence (7) and ground surveillance (8). For the defending force on the left the first priority is to use electronic support measures (ESM) to build up the picture of the enemy Electronic Order of Battle (EOB). ESM consists of airborne radar detection and analysis (9), ground monitoring (10), and sigint activities (11). Once the EOB has been established, the nature of the retaliatory action must be decided. For example, there are occasions when greater advantages will be gained from monitoring a radio net than from jamming it. In the latter case a variety of mobile jammers can be deployed (9, 12, 13) depending on the terrain, locations and frequencies involved. Alternatively, the ESM-derived information can be used to provide targets for artillery (14) or other means of fire support. Finally, it is absolutely vital that the ESM and ECM battle is carefully controlled and coordinated (15), as haphazard and uncoordinated EW will lead to chaos and a waste of effort.

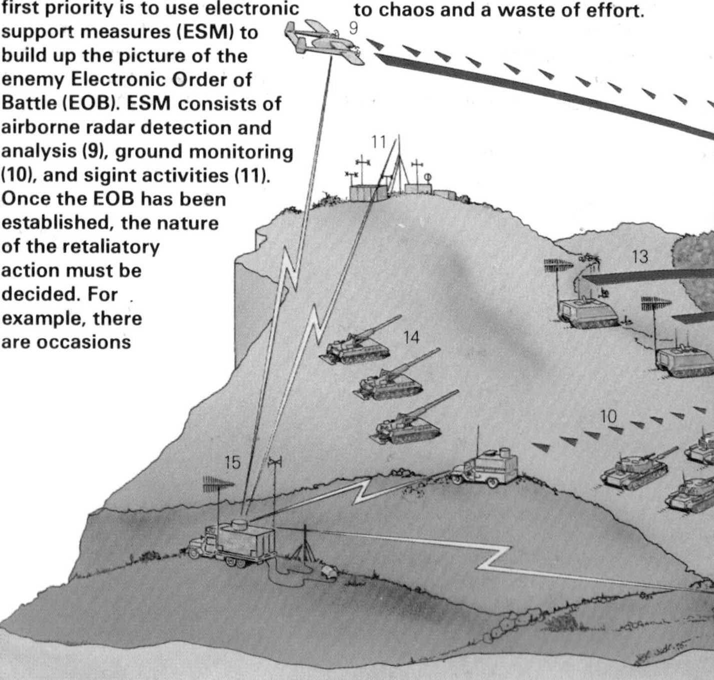

as with traditional military operations — as the loss of the *Sheffield* demonstrated.

Equipped with jamming systems unable to cope with the latest Soviet missile-guidance technology, Israeli aircrew suffered high losses in the opening days of the 1973 Middle East War. This disaster could have been avoided — the "new" techniques used by the Soviet SA-6 anti-aircraft missiles had been used by US and other Western equipment for more than 15 years! Due to an oversight in "Wizard Warfare", the jamming systems supplied by the United States were designed to cope not with such technology but with the more primitive techniques used by earlier Soviet missiles.

The days of the James Bond spy and the "mole" may not be over, but electronics also plays an ever-growing part in intelligence gathering. Antenna-festooned Soviet-bloc "trawlers" keep watch on NATO fleets, while both East and West use manned aircraft to monitor each other's radio and radar signals. American technicians have even eavesdropped on the private telephone conversations of the top men in the Kremlin.

Despite the importance of electronic warfare, the subject receives scant public attention. Many EW techniques are rarely described or discussed outside of secret research laboratories, specialized books and journals, and the military units which put them into practice. This is partly due to the tendency which some observers have of measuring the effectiveness of an armed force in terms of the numbers of "hard kill" weapons deployed. Aircraft, tanks, warships and missiles may be counted and tabulated, but the effectiveness of EW systems cannot be assessed by such simple reckoning.

Another reason is the very nature of the technical battle of countermeasure versus counter-countermeasure conducted in electronics laboratories around the world. Even the most loquacious company public relations spokesman will become tight-lipped when asked about EW — the longer that details of a newly-devised technique or equipment are withheld, the longer it may remain effective.

Unfortunately, security is not always compatible with efficiency in high-technology warfare. Barely days before hostilities began in the South Atlantic, the author was assured by a recently-retired Royal Navy officer that the service's warships were equipped with secret EW equipment so effective that Exocet could easily be countered.

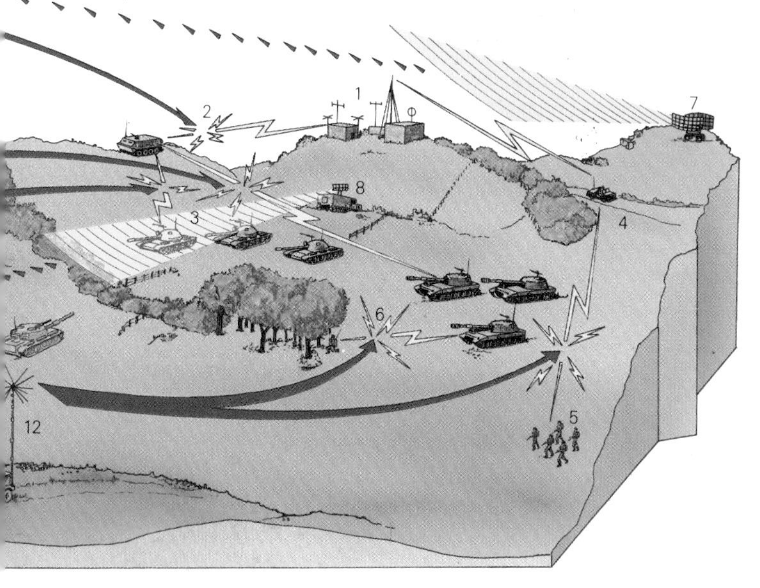

The Electromagnetic Spectrum

The battleground for electronic warfare is the electromagnetic spectrum. Like mass and gravity, electromagnetic radiation is one of the fundamental components of the universe. It encompasses a wide range of frequencies from the lowest used in long-range radio communication, through the bands used for radio and TV broadcasting and for radar, infra-red radiation, the familiar optical spectrum of visible light from red to violet, and on into ultra-violet, X-ray and gamma radiation.

The region used for military purposes normally extends from the start of the low-frequency (LF) band at 30kHz, up to 18GHz, the top end of the centimetric radar band, although recent developments exploit the extreme ends of the spectrum — very low frequencies (VLF) and millimetre waves. Reference to centimetres and millimetres may seem odd at first sight, but they refer to the wavelength of the signals involved. Engineers tend to specify transmissions in terms of frequency rather than wavelength, but the latter is often used in radar work, since it gives a direct clue to the physical size of the components used for signal transmission.

VLF transmissions — below 30 kHz — are used largely for communication with strategic submarines, and for the world-wide Omega radionavigation system. Low-frequency (LF) transmissions (30-300kHz) are used for low-quality but reliable long-range communications, and for navigation aids (navaids) such as Loran. Medium frequency (MF) — 0.3-3MHz — is the old "medium-wave" band used largely for broadcasting, although some Loran and ADF (automatic direction-finding) equipment also shares the band. High frequency (HF) — 3-30MHz —

Kara class cruiser electronics and weapons

Below: Included in this diagram of the sensors and weapons installed on a Soviet Navy Kara class ASW cruiser are the ship's EW systems. The aptly-named hemispherical radomes (14) often seen on modern Soviet vessels are for the Side Globe ESM system, and probably handle the lower frequencies, with the less prominent Bell system (11) covering the higher portion of the microwave spectrum. For sensing the bearing of HF transmissions a cross loop antenna (14) is carried.

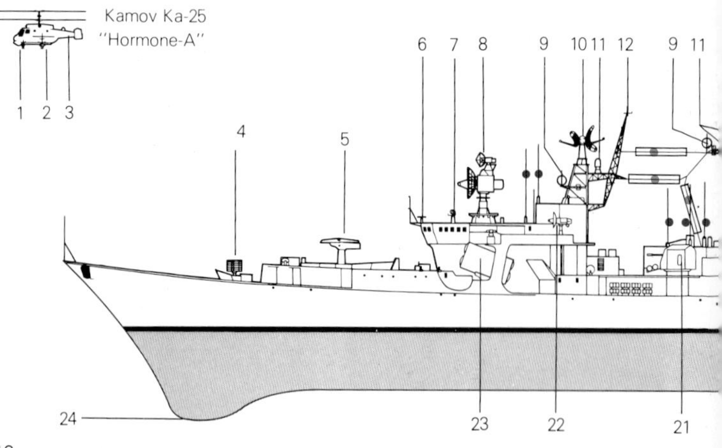

is what used to be called "short-wave". HF signals can travel long distances after being reflected by the Earth's ionosphere. Until the arrival of satellite communications, HF was the only method of long-range radio communication. It is also used by the over-the-horizon (OTH) radars deployed by both superpowers.

VHF and UHF

Very high frequency (VHF) and ultra high frequency (UHF) span the spectrum from 30-150MHz and 150-400MHz respectively. These frequencies are not reflected by the ionosphere, but are ideal for short-range communications, IFF (identification friend or foe) systems, Tacan (Tactical Air Navigation) and long-range search radars.

Above these frequencies, radio waves can no longer be passed down conventional cables, but must be led from one part of an electronic system to another using waveguides, which are metal pipes often of rectangular cross-section. This is the part of the spectrum exploited by most forms of radar.

Above: Test site for the US Navy's extremely low frequency (ELF) low data-rate strategic communications system is at Clam Lake, Wisconsin.

For convenience, the 1-20GHz area of the spectrum is divided into bands designated by letters of the alphabet between D and M. An older system of band designations, originally devised as a security measure during World War II, is still in limited use, while the Warsaw Pact uses a numerical band designation system in which these frequencies are split between bands 9 and 10.

1 Search radar
2 Dunking sonar
3 Magnetic anomaly detector
4 Anti-submarine rocket launcher
5 SA-N-3 "Goblet" SAM launcher
6 "Don-2" radar (I/J-band navigation)
7 "Don-K" radar (I/J-band navigation)
8 "Head Light B" radar group (G/H/I-band SA-N-3 missile fire control)
9 "Cross Loop" HF direction finder
10 "Head Net C" V-beam radar (E/F(?)-band long-range air surveillance)
11 "Bell" ECM
12 "High Pole" IFF transponder antenna
13 "Top Sail" radar (D(?)-band long-range 3D air surveillance)
14 "Side Globe" ESM antennas
15 "Bass Tilt" radar (30mm gun fire control)
16 Variable-depth sonar
17 Torpedo tubes
18 30mm Gatling gun CIWS
19 "Pop Group" radar (G-J(?)-band SA-N-4 missile fire control)
20 SA-N-4 SAM launcher
21 Twin 76mm gun
22 "Owl Screech" radar (G(?)-band 76mm gun fire control)
23 SS-N-14 "Silex" anti-submarine missile launcher
24 Hull-mounted sonar
• HF whip antennas
• Wire communications antennas

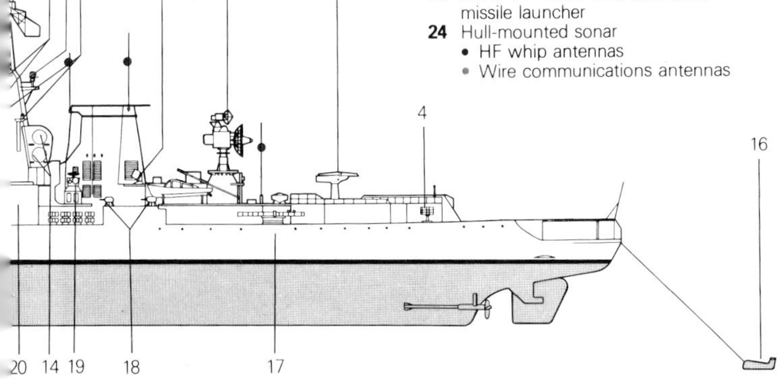

Above: One weakness of NATO is that many systems use a few overloaded radar bands. Early in the E-3 AWACS programme it was discovered that the aircraft's powerful radar transmissions could disturb the F-band surveillance radar of the British Rapier SAM.

Right: Most radars operate between 1GHz and 20GHz. This region of the spectrum is crowded with signals — only a few typical applications are shown here — placing severe demands on signal-processing equipment used to decipher them for elint, ESM and radar-warning purposes.

Below: The electromagnetic spectrum stretches from very low frequencies (VLF) to the realm of X-rays and Gamma rays. Most military activity takes place in the HF, VHF, UHF and SHF regions.

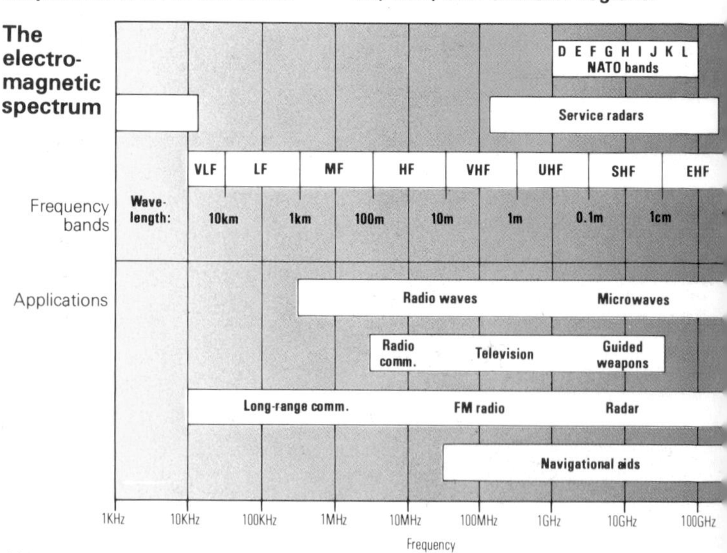

Radar applications

GHz	Band	Surveillance/tracking	Aircraft	Missile control
1				SA-3 command link (Sov)
	D	Martello 3-D surv (UK)		
		FPS-117 3-D surv (US)		
2		FPS-6 heightfinder (US)		
	E	P-50 "Barlock" GCI (Sov)		
3		RAT-31S surv (Italy)	APY-1 EW (NATO E-3)	Crotale target acquisition (Fr)
	F	RAN-10S surv (Italy)		Aegis SPY-1 multi-role (US)
				Rapier target acquisition (UK)
4				SA-2 missile control (Sov)
5	G			
				SA-6 target acquisition (Sov)
6				
7	H			
8				SA-6 target tracking (Sov)
		Rasit battlefield surv (France)		
9	I	PPS-6 battlefield surv (US)	"Skip Spin" AI (Sov Su-15 "Flagon")	SA-3 target tracking (Sov)
		TPN-18A GCA (US)	"Puff Ball" search (Sov M-4 "Bison")	
10		Cymbeline anti-mortar (UK)	APG-66/-68 multimode (NATO F-16)	
				Crotale command link (Fr)
				Seawolf target tracking (UK)
11				
			APQ-99 TF/ground map (NATO RF-4)	
12				Rapier command link (UK)
				Seawolf command link (UK)
13				
	J		"Jay Bird" AI (Sov MiG-21 "Fishbed")	
14				
15		"Gun Dish" ZSU-23-4 fire-control (Sov)		
16			APQ-113 multimode (US F-111)	
		MPQ-4 anti-mortar (US)		
17				

AI: Airborne intercept
EW: Early warning
GCA: Ground-controlled approach
GCI: Ground-controlled intercept
surv: Surveillance
TF: Terrain-following

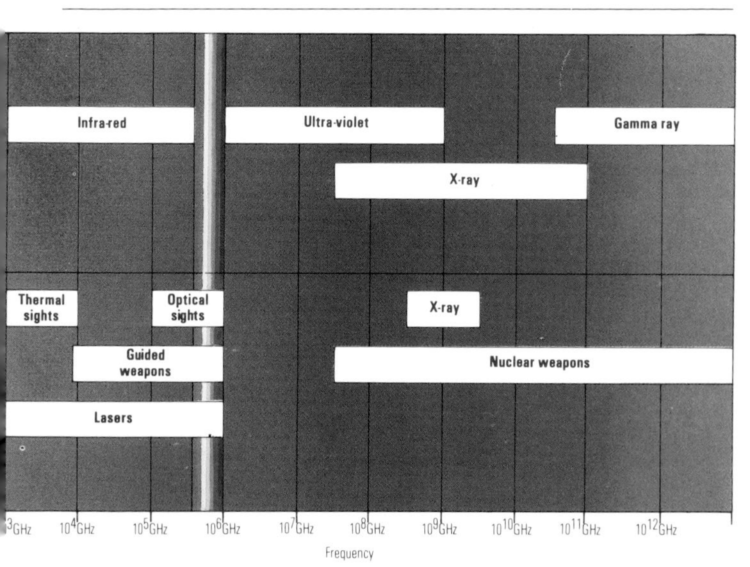

Radar

"Know your enemy" remains a valuable military maxim. In order to appreciate the world of electronic warfare, some idea of how threat systems function is necessary. If EW techniques rejoicing in exotic names such as Range Gate Stealing are to make sense, one must be able to appreciate just what the valuable gate being stolen is, and how its theft will affect the owner. A crash course in radar and electro-optical sensors is essential. The account which follows might be thought simplistic in a college of electronic warfare, but will give a basic idea of the techniques used – a road map to the mysterious world in which EW is fought.

The most commonly deployed long-range electronic sensor is radar (radio detection and ranging). Long-ranged and reliable, it is fitted to ground, ship, air and even space-based platforms. Ground-based radars are used in roles such as surveillance and early-warning, target detection and tracking, plus guidance and control of missiles or interceptors. Naval operators would expand this list to include navigation, while air forces also deploy aircraft-mounted radars for specialized functions such as ground mapping, terrain following or terrain avoidance, tail-warning, collision avoidance and altitude measurement.

Radar detects targets by illuminating them with powerful radio waves, then receiving the reflected energy. The system is crudely analogous to a searchlight, the target's only role being to reflect the energy with which it has been lit. Target bearing can be determined by noting the direction in which the antenna is pointing when returned energy is observed, but measurement of range requires a more subtle approach.

Pulse radar

The traditional method is pulse radar, which relies on the echo principle: the radar transmits a brief pulse of energy, then shuts down its transmitter and listens for an echo. The speed of radio waves, like that of light, is a constant, so by noting the time between sensing the pulse and receiving the echo, target range may be calculated. Once enough time has been allowed for the pulse to have reached the distance which the designer had adopted as the maximum range of his radar, a second pulse may be sent and a new range determined.

The rate at which pulses are sent is called the Pulse Repetition Frequency (PRF). Long-range radars have low PRFs, while short-range radars have high PRFs. Designers often provide several PRFs, so that a higher value may be used when operating over shorter ranges.

The ability of a pulse radar to discriminate between several targets appearing on the same bearing is dependent on the length of the radar pulses. If the pulse length is longer than the time needed for the pulse to travel from one target to another, then the echoes from these targets will overlap and the radar will be

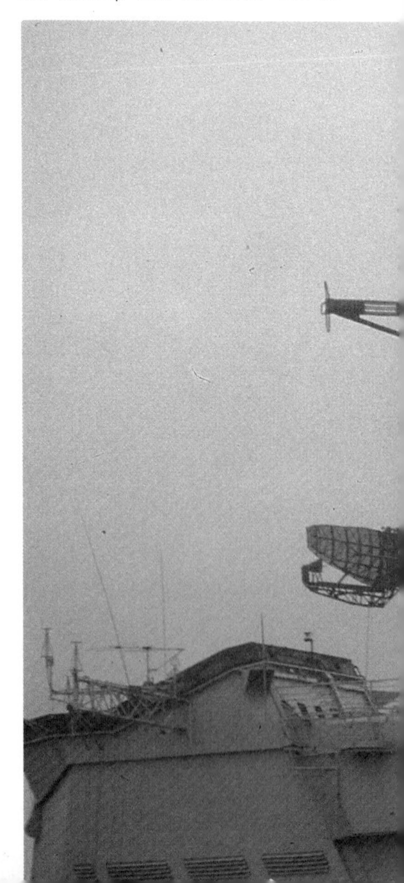

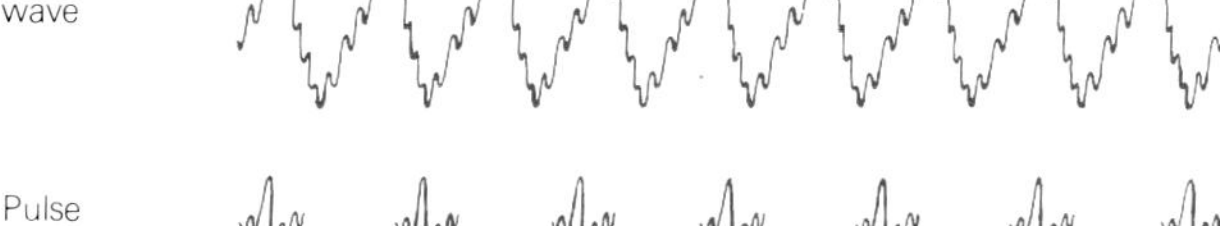

unable to distinguish between them.

The obvious solution is to make the pulse length as short as possible, but this creates a further problem. As the pulse length decreases, so also does the energy contained in each pulse. To some degree, this dropping off in power can be compensated for by increasing the peak power of the pulse, but transmitter technology places limits on the degree to which this may be done.

One compromise often adopted by radar designers is to provide several pulse lengths. For long-range search a relatively long pulse length may be used in order to maximize power output, with a shorter pulse being used for maximum discrimination at

Above: Two patterns of radar signal are in common use — continuous wave and pulsed.

Below: The spherical radome at the masthead of the Soviet carrier *Kiev* is for the "Top Globe" missile guidance radar, while the circular antennas of the "Head Light" radar (far right) are for the SA-N-3 "Goblet" missile. Forward (right) of the main mast is the massive lattice antenna of the "Top Sail" air-surveillance radar, while the back-to-back "Top Steer" antennas are located aft. Several EW antennas (probably for ESM) are visible at bottom centre.

shorter ranges. Like PRF, pulse length (often known as pulse width) is a valuable "fingerprint" in identifying an enemy radar signal.

Above: Perhaps the Israelis can explain how this Soviet P-50 "Bar Lock" early-warning radar ended up on a US test range. The large width of the antennas gives good azimuth accuracy.

Continuous-wave radar

A second type of radar used for military applications does not rely on pulse operation for ranging. Since the transmitter radiates continuously, such sets are known as continuous-wave (CW) radars. In order to obtain range data, a CW set operates not on a fixed frequency but over a band of frequencies. The output of the transmitter is varied with respect to time. By measuring the frequency of the return from a target it is possible to calculate the time interval that has elapsed since a signal at that frequency was transmitted, and thus to calculate target range.

Antennas (aerials)

Radar antennas usually take the form of a solid or mesh reflector illuminated by a horn or other source of energy. They operate on the same principle as a spotlight, with a single source of energy illuminating the reflector in order to form a beam.

The antenna of a radar is designed to produce the beam shape best suited to the task for which the set was designed. A slim pencil-like beam is best for target tracking, while semi-focused fan-shaped beams are often used for search purposes. Mechanical changes to the antenna (such as the use of "spoilers") allow a modest measure of beam shaping for multi-role applications, but a conventional antenna is basically designed to produce a single type of beam customized to a greater or lesser extent to match the operational requirements which the set must meet.

In addition to the main beam, antennas also have smaller unwanted beams (known as sidelobes) which radiate at odd angles from the main beam or even directly to the rear. Designers attempt to minimize the size of sidelobes, since they can be and often are exploited by hostile EW systems.

Surveillance radars

Surveillance radars are usually intended to give accurate range and bearing information while coping with targets at varying elevation angles, so tend to use a fan-shaped beam no more than a few degrees wide but 30° or more in height. Since

Right: The antenna shape of heightfinding radars such as this Marconi S613 gives a wide coverage in azimuth, but optimum discrimination in elevation between individual targets.

beamwidth is inversely proportional to antenna dimensions, antennas used on sets of this type tend to be wide but relatively low. Specialized heightfinding radars used to measure the altitude of aircraft targets detected by surveillance sets require first-class elevation accuracy, so tend to be narrow but very tall. Tracking radars must have good accuracy in both planes, so need antennas whose height and width are similar.

The scanning pattern is also dependent on role. Surveillance sets tend to scan a sector or a full 360°. Once turned onto the bearing of a target, heightfinders have a characteristic "nodding" scan pattern unique to radars of this type. When operating in search mode the radars carried by fighters often search for targets by using a multi-bar raster scan pattern. Once a target has been located they will, like most radar trackers, keep their antenna pointing towards the target with a fixed or near-fixed "gaze".

Frequency-hopping radars

Although these basic principles have remained unchanged for decades, the military effectiveness of radar has been greatly increased by more advanced technology. Early sets worked on a stable fixed frequency, which was either pre-set at time of manufacture or could be selected in the field from several options. Such simple equipment was relatively easy to monitor or jam, so frequency agility is a feature of many modern radars: the operating frequencies of the transmitter and receiver are rapidly switched in an unpredictable manner over a band of values, making it difficult for search receivers to locate the signal, and rendering useless the simple jamming equipment designed for use on a single pre-set frequency.

Frequency agility offers a further bonus. An aircraft or ship may seem of near-constant size to the eye, but to radar will often fluctuate in apparent size. At any one frequency, target size is very dependent on target attitude, so can alter very rapidly as small changes in target aspect cause variations in the amount of energy reflected. Frequency agility reduces this problem. At any given target attitude, pulses on some frequencies will be strongly returned, while others at frequencies less well suited to the target will be returned more weakly. As the transmitter frequency changes from pulse to pulse, so does the strength of the received signals. By integrating the amplitude of a large number of successive pulses, the radar can largely eliminate the effects of such fluctuations.

Travelling-wave tubes

One problem in creating frequency-agile radars was the fact that the cavity magnetron – the traditional generator of microwave radar energy since the early 1940s – must be optimized for use at or close to a fixed frequency. In the search for greater agility, designers turned their attention to an alternative signal source in the form of the travelling-wave tube (TWT), which is able to operate at high power levels over a bandwidth which can typically be of up to 10 per cent of the centre frequency.

Another advantage associated with TWTs is their ability to cope with complex modulation patterns (methods of shaping the signal or adding data to it). The magnetron was ideally suited to pulse radar applications, since its duty cycle is typically around 0.1 per cent – for 99.9 per cent of the transmitting time, no power is being radiated, the entire output taking the form of a series of short but high-energy pulses. This very high on/off cycle is just what is needed for simple pulse radars, but rules out the use of more complex radar operating modes required for look-down applications and to give improved resistance to jamming. Once again the TWT provided useful, its duty cycle being up to an order of magnitude higher than that of a magnetron.

Analogue versus digital electronic systems

Until the last decade or so, signal processing within radar sets was entirely analogue, with targets and other data being represented by electrical signals which could be amplified, shaped or otherwise processed as required. Analogue electronics are simple and well-proven, but suffer the disadvantage of adding unwanted noise to the signals being handled, and more complex electronics mean more noise.

F-16 radar range and field of view

Below: The F-16's I/J-band pulse-Doppler radar (APG-68 in the F-16C) combines TWT power source, flat-plate planar array antenna, multiple PRFs and digital signal processing to provide a range of tactical options. In air search mode 1-, 2- and 4-bar scan patterns are available, while the 20x20° and 10x40° air combat patterns can be slewed to left or right.

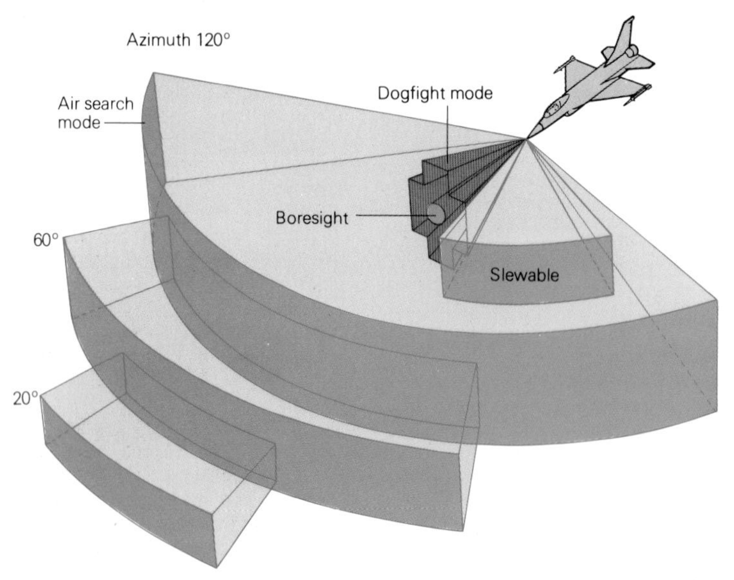

Above: Improved technology enables modern fighter radars, such as the APG-65 installed in this F-18 Hornet, to be made small and compact.

The latest generation of radars employ digital signal processing, the radar data being converted into electronically-coded numerical form in which it can be manipulated, catalogued or otherwise processed without the risk of degeneration. Within a digital electronic system, this numerical data is usually stored as binary-coded characters — a series of electrical signals which are either "on" or "off". Individual circuit elements are either "on" or "off", and are almost immune to noise or interference no matter how often the data is processed.

On a conventional analogue radar, interpretation of the display is a skill which operators must master, learning how to pick out targets from among the clutter and noise. This was particularly a problem with radars attempting to follow low-flying targets. The quality of the image on the radar display will degenerate sharply as unwanted clutter from terrain below the target starts to swamp the radar.

On the current digital radars, the display uses not the analogue radar signal (often referred to as "raw" data), but digitally-generated symbology. Instead of being shapeless splashes of light, targets are presented in symbolic form to meet the chosen preference of the user — "friendly" targets might appear as circles, "unknowns" as squares and "hostiles" as triangles, all neatly annotated by track numbers and pertinent data.

Digital data may easily be sent over long-range and often noisy communications links, techniques having been devised to identify and correct any garbles introduced by bad transmission. This facilitates large-scale data transmission, making it easy for early-warning aircraft such as the E-3A Sentry AWACS to pass detailed target information to ground sites, other AEW aircraft or interceptors. Information can be analyzed, sorted and re-distributed by computer-controlled systems without human intervention.

Pulse-Doppler radars

A combination of digital signal processing and TWT transmitters made possible the creation of pulse-Doppler radars, which are able to operate in "look-down" mode from interceptors and to track low-flying targets. By sensing the frequency shift in the target echo caused by the Doppler effect, a radar can now discriminate between radar returns from the target and the very much stronger echo from the terrain background.

Individual output pulses from the cavity magnetron bore no relationship in phase to one another, but the TWT allowed low-power signals from an ultra-stable oscillator to be used to trigger a series of coherent (phase-related) output pulses whose returned energy could be accurately compared to detect the difference in frequency.

The properties of coherent signals are difficult to put into simple terms, but a crude analogy might help. Back in the days of early rock music, the police force of one Scottish town

decided to curb the occasionally turbulent behaviour of some of the fans. Kitted out in long jackets and drainpipe trousers and with their hair suitably greased, the officers set out to mingle with the local youths, who were known as "Teddy Boys". But the tactic failed miserably: the sight of several tall "Teds" striding towards the local dance hall in precision step merely reduced the local populace to laughter. Coherent signals are as easy to spot amongst normal signals as the police "Teds" were at the dance. Their rigidly controlled behaviour marks them out.

One problem in the design of pulse-Doppler radars is that TWTs cannot match the power output of magnetrons, so higher PRFs must be used to ensure that the target is illuminated with sufficient energy. Magnetrons operate most efficiently at low PRFs (less than 5kHz), but the TWT allows the use of medium or even high PRFs. Since these higher PRFs do not allow sufficient time for

Above: Digital data processing allows modern radars such as this Thomson-CSF TRS-2230 3D to present the operator with clean, clutter-free, synthetically generated symbology.

Right: Digital technology also allows the creation of high-resolution synthetic-aperture radars. In this demonstration, an F-15's APG-63 radar maps the airfield shown in the photograph.

the individual pulses to complete the trip to and back from the target before the next pulse is despatched, the individual pulses must be modulated at low frequency so that the radar can determine which pulse caused which return, and thus calculate target range.

This calculated range is not as accurate as that obtained by low-PRF radars, so in recent years radar designers have begun to use medium PRFs in the region of 6-16kHz. Since

PRFs suitable for obtaining good range information (low enough to permit individual pulses to complete the round trip before the next is despatched) may not be the best for measuring target velocity, a series of medium-range PRFs are often used in rapid succession.

Another problem with high PRFs is experienced when the target and radar have a low velocity with respect to one another – a situation which can easily arise if a fighter is approaching its target from the rear. All-aspect and all-altitude tracking of moving targets by pulse-Doppler radar requires a range of waveforms.

Before the arrival of digital signal processing, Doppler information was derived from a series of as many as 1,000 filters in the radar hardware. These had to be designed for a given set of conditions, so PRF switching was not possible. Additional frequencies would have needed additional sets of filters. In a modern set these filtering operations are carried out by software, so can be automatically changed to match the waveform being transmitted.

Flat-plate antennas

Another feature of recent radars is the widespread use of flat-plate antennas. Instead of using passive reflectors, these consist of arrays made up from large numbers of elements known as phase shifters. Each phase shifter transmits a tiny portion of the signal, but delayed by a programmable amount designed to create a beam.

In a conventional radar the antenna must be pointed in the direction of the target. Many phased array sets use the new flat arrays as a substitute for the conventional "dish" or "orange peel" antenna, and thus must retain the traditional servo system or scanning mechanism used for antenna steering. In order to track multiple targets — a common military requirement — the antenna must scan a large pre-set volume of sky or terrain, building up a "track file" of targets from the position and velocity data obtained as each target is briefly illuminated by the scan pattern.

Alternatively, by altering the degree of phase shift generated by each element in a phased array, the radar designer can arrange for the beam to be steered or shaped as required. The antenna may remain fixed, the beam being electronically scanned to align it with the target. When tracking multiple targets, the

Right: This steerable planar array antenna is part of the terminal homing system of the AIM-120A AMRAAM (Advanced Medium Range Air-to-Air Missile).

Left: By varying the phase of the signals fed to each circular element of this antenna, the beam generated by the complete array may be shaped as required, or even steered electronically at high speed from target to target.

Below: By building large, electronically-steered phased-array antennas engineers can create blast-resistant radars able to track ballistic missiles and re-entry vehicles.

antenna can switch rapidly from one to another within microseconds, allowing all to be monitored near-simultaneously.

Electronic noise

One fundamental problem which haunts designers and operators of radar, sonar or any type of remote-sensing system is that of "noise". This may take the form of unwanted signals reaching the system via its normal input, or electronic activity generated within the electronic system itself. The amount of noise generated within the system can be minimized by good design, but never fully eliminated.

The electrons within an electronic component or even a piece of wire will move at random by an amount dependent on the temperature of the component or wire. This is a fundamental fact of physics. These movements result in tiny and random electric currents which the system will accept as a low-intensity signal.

This is no theoretical problem, as operation of some items of domestic electronic equipment will demonstrate. If a portable TV set with an indoor antenna is used in an area of weak signal, the picture will be weak and partly obscured by random, fast-moving white dots which TV technicians nickname "snow". The signal is so weak that the set's electronics are attempting to interpret noise as genuine signal. The noise is random, so the end result is random tiny patches of interference scattered over the picture area.

Noise is a significant factor in EW. Many methods of EW attack are designed to inject noise within the enemy's sensor, while some of the design techniques used by radar and sonar designers in an attempt to reduce the effects of noise – including the Range Gate concept mentioned earlier – provide a weakness which the countermeasures designer can exploit in the electronic battle of wits.

Millimetre Waves

Above 20GHz begins the region commonly referred to as millimetre waves, most of which is covered by the 30-300GHz portion of the spectrum known as EHF (extremely high frequency). NATO band designations for this part of the spectrum are K, L and M band, while the Soviet designation is 11 band.

Millimetre-wave techniques have been used in the laboratory for many years, particularly for investigating the effect of aircraft shape on the performance of antennas. By using millimetre waves as miniaturized versions of centimetric transmissions, engineers could make their trials and measurements indoors using easily handled scale models rather than outdoors using full-size aircraft and antennas. For such applications, engineers were content to use relatively crude techniques for signal generation and detection: the breakthrough in operational applications came with the development of solid-state devices able to handle such high frequencies.

Below: The Ka-band secondary channel of the tracking radar fitted to the Krauss Maffei CA1 anti-aircraft tank gives the circular antenna a narrow beamwidth of less than a 1°.

Millimetre waves are already being exploited for military purposes, and will play a growing role in future weapon systems: British exploitation of centimetric wavelengths gave the Allies an unshakable lead in the radar field during World War II, and the coming move to millimetric wavelengths could hold equal promise. This region is largely unexploited, but could be used in future fire-and-forget missiles, tactical radars, short-range communications links, radar, and weapon fuzing. Most will be short-range systems, for reasons discussed below.

Millimetric-wave operating frequencies allow high-gain/high resolution antennas of modest dimensions to be created. The signal beam can be made very narrow, and thus difficult to detect and monitor. Sidelobes — spurious secondary beams which reduce system performance and act as an Achilles' heel

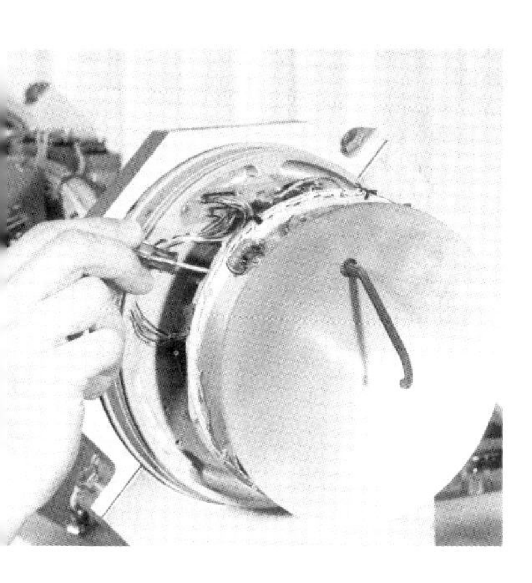

Above: The miniature waveguide which feeds the antenna of this experimental Hughes missile seeker shows how small millimetre-wave components can be.

which enemy ECM designers will attempt to exploit — are minimized.

One potential problem is atmospheric attenuation resulting from absorption of the signal by atmospheric oxygen or water vapour. For applications which require long range, millimetre-wave equipment must operate at frequencies where these effects are only minimal — regions (referred to as "windows") centred around 35, 94, 140 and 220GHz. The degree of attenuation is less than that suffered by infra-red radiation, so millimetric sensors are less affected by smoke, fog and rain than IR systems, but they do have a lower resolution for the same sensor size, because of their lower operating frequency.

Applications

Millimetre-wave transmissions are able to carry a large amount of data: any one of the windows described above offers a greater bandwidth than all the conventional radio/radar spectrum put together. This can create problems when searching for hostile signals during sigint operations.

One key application of millimetre-wave technology is in the field of radiometry. Millimetre waves are almost a half-way house between the world of radar and that of infra-red: in the same way that all objects radiate heat, they also radiate millimetre waves, and the intensity of such radiation is proportional to what engineers term the radiometric or noise temperature of the object in question. This is not necessarily related to thermal temperature, although it often is. By exploiting this millimetric radiation, designers can create passive seekers or even dual mode active/passive millimetre-wave seekers. Other patterns of dual-mode sensor may combine active or passive millimetric modes with laser or IR homing.

Although the theoretical upper limit of the millimetric portion of the spectrum is 300GHz, a practical upper limit for the use of conventional components such as transmission lines and waveguides is around 200GHz. Beyond this point signals must be treated more like infra-red or light, rather than by radio/radar techniques.

Below: The narrow beam from the antenna of this Norden millimetre-wave radio makes tactical communications difficult to intercept by any unauthorized eavesdropper.

Infra-red

All objects radiate infra-red (IR) energy, and the hotter the object, the greater the energy emitted. The infra-red band covers wavelengths of 1.5-14 microns; the IR spectrum for a warm object occupies a broad band of frequencies within this range. The peak of the emission pattern of the emitted energy depends on its temperature – the hotter the object, the shorter the wavelength.

Not all the IR spectrum is militarily useful, however, since here again water vapour, carbon dioxide and other constituents of the atmosphere absorb energy in many regions of this part of the spectrum. As a result, IR system designers tend to use one of two "windows" in the spectrum. 2-3 microns and 8-14 microns. Objects heated to around 300°K (27°C, 81°F) have emission peaks in the 8-14 micron band, a rise in temperature to around 700-800°K (427-527°C, 800-980°F) is needed to create a peak at 2-3 microns.

Heat-seeking missiles

The main military applications of infra-red technology are imaging and non-imaging seekers, and night vision systems. Widely used in heat-seeking anti-aircraft missiles, non-imaging seekers are normally made in the form of a small telescope designed to focus the incoming infra-red energy onto a sensor consisting of one or more detecting elements. These are often cooled to very low operating temperatures in order to increase their sensitivity.

The first infra-red guided weapons such as the AIM-9 Sidewinder air-to-

Below: The extended tailpipe of this Israeli A-4 Skyhawk is a counter to IR-guided SA-7 "Grail" missiles. During the 1973 Middle East War, large numbers of A-4s were grounded with damaged tail sections, but this low-cost add-on now absorbs most of the blast from any SA-7 detonation.

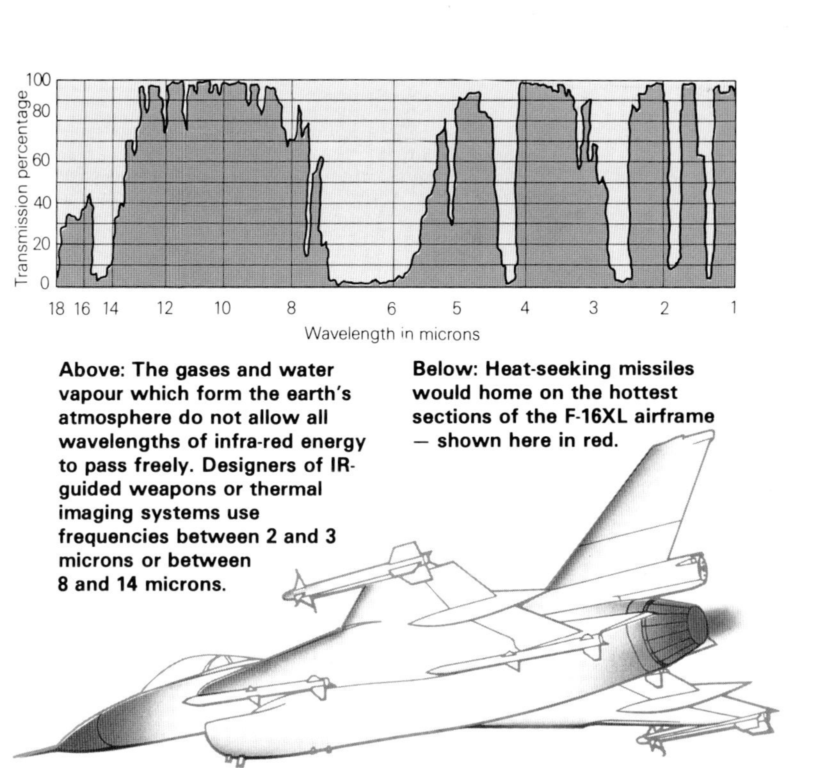

Above: The gases and water vapour which form the earth's atmosphere do not allow all wavelengths of infra-red energy to pass freely. Designers of IR-guided weapons or thermal imaging systems use frequencies between 2 and 3 microns or between 8 and 14 microns.

Below: Heat-seeking missiles would home on the hottest sections of the F-16XL airframe — shown here in red.

air missile entered service in the early 1950s, and had seekers with a fixed forward-looking gaze. In combat, the pilot of the launch aircraft had to manoeuvre so as to bring the intended victim into the seeker's field of view – a narrow cone only a few degrees wide. Seekers were of low sensitivity, so could respond only to the hot rear section of the target's engines, flying virtually up the jetpipe of their victims.

In modern designs the seeker is steerable, and may be programmed to carry out a pre-launch search procedure covering a selected area, or slewed onto the target bearing in response to signals from the parent aircraft's radar or other sensors. Greater sensitivity also allows the seeker to respond to the lower intensity heat radiated by most if not all of the target's airframe, so launch is possible at virtually any angle.

Non-imaging seekers can cope only with a single hot source against a fairly uniform background: such simple technology lacks the sophistication needed to analyze a complex infra-red image. As a result, they are useful only for anti-aircraft and anti-ship missions – instances when the target may be expected to be much hotter than the background. When simple IR-guided missiles such as the Soviet AA-2 "Atoll" and early-model Sidewinders attempt to lock on to targets flying lower than the launch aircraft and seen against a relatively hot background such as desert terrain, their seekers are easily confused, and can easily miss the target.

Imaging infra-red

The fire-and-forget potential of IR guidance makes it attractive for air-to-ground use over the battlefield, but this requires the use of more complex seekers based on imaging infra-red (IIR) technology. These seekers are virtually infra-red TV cameras which build up a heat image of the target and background, then rely on sophisticated signal processing to lock on to a designated part of the image, such as one edge or a bright or dark feature of the target. Unlike TV systems, IIR sensors work equally well in total darkness, since targets do not lose their heat by night. They are also better than visual systems at coping with haze and smoke, although since fog and cloud consist of suspended water vapour which attenuates IR, these conditions remain troublesome.

IR systems will in the future supplement or even replace radar-based target-detection systems. A passive IR sensor able to detect incoming anti-ship missiles was developed under the US/Canadian Infra-red Search and Track (IRST) programme. The resulting SAR-8 equipment underwent development trials aboard the warships HMCS *Algonquin* and USS *Kinkaid* in the late 1970s. Further work was to have resulted in the Texas Instruments Seafire FLIR (forward-looking infra-red) based naval fire-control system); this was later cancelled, but work is under way on a new system.

In the UK, British Aerospace is working on a passive version of the Rapier SAM. Tracked Rapier vehicles of the British Army are to be updated with the TOTE (Tracker Optical Thermally Enhanced) system, a modified tracker which allows target and missile to be followed using optical or IR viewing systems. According to reports, the planned Field Standard C Towed Rapier for the British Army will be largely passive in operation, using a new tracker based on a thermal imaging system rather than conventional optics, and a PASS (Passive Aircraft Surveillance System) thermal detector and a 5MW CO_2 pulsed laser ranger instead of a surveillance radar.

IIR technology has also been applied to night-vision systems. Primitive night-vision systems relying on the use of infra-red searchlights have been in service for decades, but these simply converted the infra-red scene lit by the searchlight. Once the enemy had deployed IR viewing systems, using these searchlight-based equipments was little more subtle than shining a conventional searchlight. Modern thermal imaging systems are passive, relying on the heat emitted by objects in the field of view. Now widely deployed as surveillance systems, navigation aids and weapon sights, they have proved reliable and efficient.

Right: In these infra-red images of the 65t Conqueror main battle tank, hot parts of the vehicle appear white. Conqueror is obsolescent and no longer in British Army service — IR images of more recent tanks remain classified.

Below: These IR images of the USS *Bagley* were taken during trials of the imaging infra-red seeker for the AGM-65F Maverick. The top image was obtained at the moment of lock-on, well beyond visual range, the others at shorter distances.

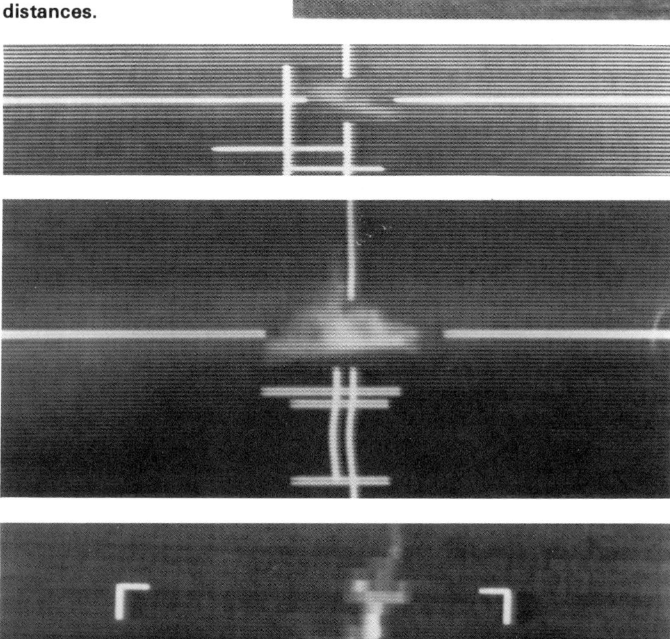

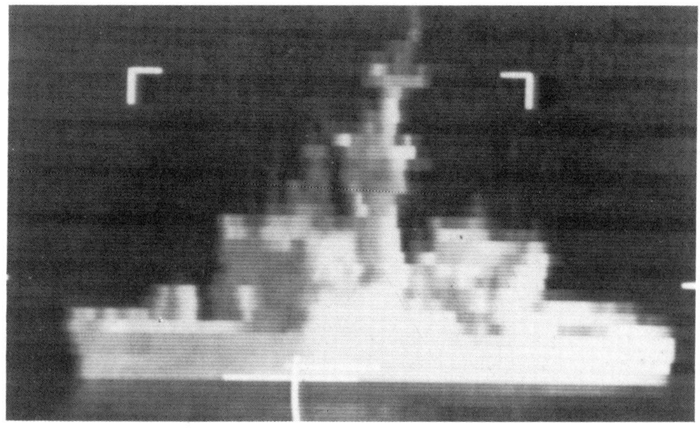

Electro-optical Systems

For a long time the primary military sensor was the human eye, helped by binoculars and telescopes, but electronics has allowed the development of electro-optical (EO) sensors and seekers operating in the near infra-red and visible light portions of the spectrum. Electro-optically guided weapons first saw service with the US forces in the Vietnam War, and the latest generation — systems such as the Maverick TV-guided missile and Walleye smart bomb — are widely deployed. These carry a miniature TV sensor in the nose, and may be locked onto a suitable point of high contrast on the target. Similar EO seekers are also used as target trackers.

Like image-intensification night sights and viewing systems (which magnify the tiny amount of light available from the sky on even the darkest night), EO seekers are passive, so do not warn the enemy that they are being used. But of course the main problem is that of visibility, particularly in typical Western European weather conditions, so many guided weapons now in development and entering service tend for this reason to use imaging infra-red technology.

One novel future application of electro-optics is the seeker of the planned man-portable anti-aircraft missile being developed by the Japan Self-Defence Force Technical Institute. Rather than use IR homing, the Japanese weapon will lock on to the visual image of the target, allowing all-aspect engagements.

Below right: Despite NATO's optimistic studies, attrition in any Central Front war would make the disastrous losses which Israel experienced in the opening days of the 1973 War seem mild by comparison. Improved accuracy in air attacks is vital, and smart weapons like these TV-guided Mavericks, seen with an A-10 which would be a principal delivery vehicle, would make every sortie count.

Below: A USAF F-4 Phantom releases a Maverick TV-guided missile. The application of electro-optical homing to smart weaponry has triggered off the development of countermeasures operating in the visible part of the electromagnetic spectrum.

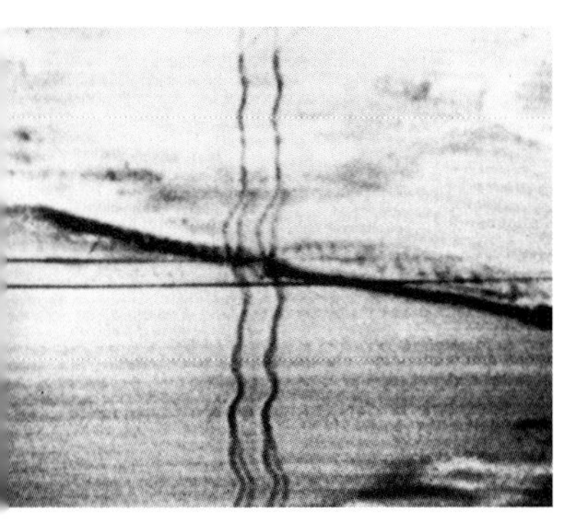

Above left, above and left: TV images transmitted by the weapon's seeker recorded the last moments of a 1972 Walleye attack on the Ninh Binh railroad/highway bridge in North Vietnam. Once locked on to the high-contrast portion of the image marked by the crosshairs (above left), the weapon homed in to impact, continuing to transmit TV pictures to the launch aircraft.

Sonar

Radar signals have virtually no penetration into water, so an alternative long-range sensor is needed for underwater search. Sonar (sound navigation and ranging) is the acoustic equivalent of radar. In its active form, electrical signals are converted into pulses of ultrasonic sound by a transducer and used to locate submarine targets. Passive sonar involves simply listening with directional receiving equipment, locating targets by means of the sounds made by their machinery and equipment, or even their passage through the water.

Passive sonar yields only bearing information. In order to obtain range, the sonar platform must move to another position in order to create a baseline for triangulation. The target itself is moving, and may change course. If a change of heading such as a zig-zag manoeuvre is not detected and allowed for, the derived range will be wrong. As range decreases, the usefulness of triangulation by passive sonar is greatly reduced.

This problem is particularly acute in submarine versus submarine actions. At the very time when good target data is required, the attacker is reduced to receiving only bearing data. In order to attack, it must either switch to active sonar, or rely on firing homing torpedoes down the target bearing. Once the attacker starts "pinging" its transmissions will alert its intended victim, which

Right: The left-hand display on the console of this EDO variable-depth sonar shows the predicted transmission path of the sonic pulses from the towed transducer. The right-hand screen displays information on the targets detected.

Below left: Under good propagation conditions, high-definition sonars can produce spectacular images. A US Navy side-scanning sonar located this relic from the past — a B-25 Mitchell bomber on a South Carolina lake bed.

Below: A warship prepares to deploy its variable-depth sonar. The towed "fish" seen here will carry the sensor to the depth giving the best results in the prevailing water conditions.

may counter by launching a long-range anti-submarine weapon of its own.

Several methods of avoiding this problem have been postulated, but none shows signs of being practical. Use of a remotely-positioned sonar transmitter would hide the true bearing of the attacking submarine, but would still warn the victim. Some form of covert sonar with difficult-to-detect transmissions is theoretically possible, but the practical development problems remain formidable.

The water is not a good medium in which to propagate signals. Radar signals move through the air in straight lines, but the transmission path of sonar signals in water may be twisted and distorted by changes in water temperature or salinity. Layers of water through which sonar signals will not easily penetrate exist at various depths, and may be exploited by a submarine trying to avoid detection, so some sonar sets have sensing elements located inside a housing which can be lowered into the water and under the troublesome layers.

Sonar platforms

Sonar systems may be mounted in surface vessels or submarines, dropped from fixed- or rotary-winged aircraft, "dunked" from helicopters, or pre-positioned on the seabed. Like latest-generation radars, modern sonars use digital signal processing techniques. By retrofitting existing transducer arrays with modern electronics, navies can obtain a significant increase in performance at moderate cost.

Most surface ships are fitted with sonars. These may be hull-mounted systems with sensing elements within domes mounted under the bows or keel, or the variable depth systems mentioned earlier. Another way of getting a sonar sensor deep into the water is to mount it aboard a submarine. Nuclear powered attack submarines are highly efficient hunter-killers, and outnumber their ballistic-missile-armed counterparts. The US Navy no longer builds conventionally-powered submarines, and the 30+kt underwater speed of its *Los Angeles* class boats allows these vessels to operate in direct support of carrier groups.

Above and above right: Once in the water, the Plessey Cormorant — dunking element of the HISOS 1 helicopter-mounted sonar — deploys sensor-equipped arrays and begins to search for targets.

Towed arrays

The increased range of the missiles and torpedoes now being carried by Soviet submarines has forced Western navies to supplement active

sonars with towed passive-array equipments intended to detect targets at long range. This type of equipment first saw service aboard submarines, but larger and more elaborate systems have since appeared aboard surface vessels.

The sensor elements consist of long lines of neutrally buoyant cable towed behind the sonar equipped vessel and carrying a series of low-frequency hydrophones. The beam-width of the array is dependent on operating frequency, array length, and the angle at which sounds reach the array. Typical figures might be around 2° for targets on the ship's beam to 5° near the bow or stern.

Sonobuoys

Maritime patrol aircraft such as the US P-3 Orion, British Nimrod, French Atlantic and Soviet Il-38 "May" must drop sonobuoys in order to get sensors into the water. Although expendable, these are used in large quantities, with individual manufacturers delivering tens of thousands per annum. On landing in the water, a sonobuoy raises its

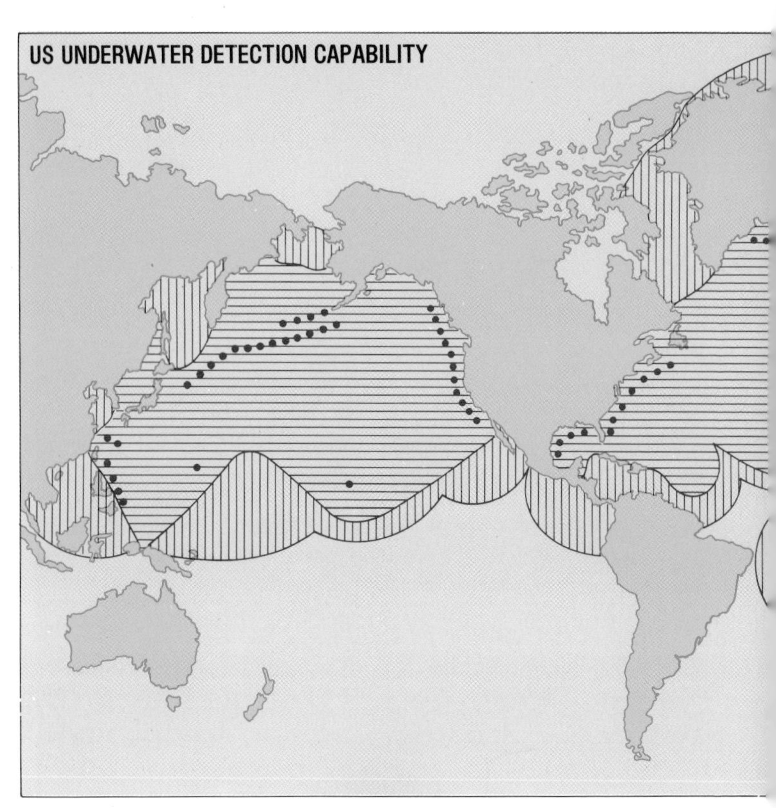
US UNDERWATER DETECTION CAPABILITY

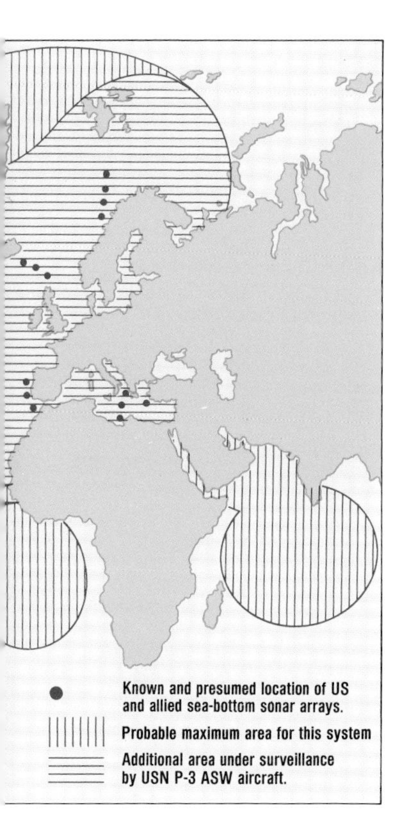

Above: A combination of sea-bottom sonar arrays and patrols by ASW aircraft gives the US Navy good coverage of much of the world's oceans. Just how reliably individual Soviet submarines may be detected and tracked remains classified.

Left: Most details of the US seabed arrays remain classified. This photo shows the lowering of an "antenna array barge" near Andros Island, Bahamas. In US defence budgets, the costs of such work are often charged to "hydrographic research".

antenna and releases a sensor, which falls to the end of a cable whose length may often be preset before the unit is dropped. Operating life is fixed or pre-selected before release, the buoy scuttling automatically once this time limit has expired.

Sonobuoys may be directional or omnidirectional, and like sonars may operate on active or passive principles. The simplest buoys are passive and omnidirectional, but the most sophisticated units such as the Dowty Electronics CAMBS (Command Active Multibeam Sonobuoy) are active and directional. CAMBS can vary its sensor depth on radio command, giving the aircraft which released it much of the flexibility normally associated with dunking sonars fitted to many types of ASW helicopter. Dunking sonars can complement the area surveillance capability of sonobuoys, and when used in passive mode can provide a covert surveillance method.

Seabed sonars

The longest-ranged sonars are the fixed systems located on the seabed. Best-known of these is the US Navy's Sound Surveillance System (SOSUS), intended to detect and track submarines operating off the coasts of the United States, or transiting choke-points in the North Atlantic and Baltic and near the Azores. The original system – then known as CAESAR – was laid on the continental shelf off the New England coast, becoming operational in the early 1960s. This was followed by a complementary west coast network, while additional systems with designations such as Bronco, Barrier, Colossus and Sea Spider were deployed in the maritime areas of NATO allies and near Hawaii and other strategic islands.

Effectiveness of the system is classified, but is sufficient to detect and roughly locate most patterns of submarine. Once the hydrophone network has distinguished the acoustic signature of a target submarine from ambient noise, a land-based ASW aircraft such as a P-3 or Nimrod is used to locate the vessel accurately and commence tracking.

In order to plug gaps in the coverage due to failure or combat damage, or to monitor areas of sea not under permanent underwater surveillance, the US Navy devised the shipborne Tagos SURTASS (Surveillance Towed-Array Sonar System), and is developing the air-dropped Rapidly-Deployable Surveillance System (RDSS).

Stealth Technology

One new factor in the electronic battle between aircraft and anti-aircraft weapons and sensors is the use of stealth technology. Highly classified by the US Government, this is probably the most secret type of military hardware. Photographs have been released of current US nuclear weapons, but all stealth aircraft and missiles have been kept under wraps, and are likely to remain so until operational hardware is deployed overseas.

In some company plants, stealth-related work is carried out in restricted areas off-limits to most employees. Early leaks of information were ruthlessly suppressed on the orders of the US Government, while a policy of retrospective classification was applied to openly-published documents and technical papers which had described technologies applicable to stealth aircraft.

Despite these draconian measures, some information has leaked, and it is possible to describe the main lines of current development. It is clear that stealth involves not a single technology, but a blending of several techniques — careful shaping of the airframe, the use of radar-absorbing materials, and the incorporation of sophisticated ECM techniques. None of these techniques is new, but used in combination their effectiveness is greatly multiplied.

Non-reflective materials

Best-known method of reducing radar target size is the use of non-metallic structural materials. Radar relies on the fact that when a radar signal strikes an aircraft it induces electrical currents within the structure, and that these in turn emit the energy which will form the radar echo. If a non-conducting material is used, very few currents will be induced, and the resulting echo will be minimal. Non-metallic substances such as carbon composites, glass fibre and even more exotic com-

Below: The fourth Rockwell International B-1A prototype was the first to carry a camouflage finish. This view shows one radar-reflecting feature deleted from the B-1B — the long dorsal strake originally designed as an avionics housing.

Above: The fact that publication of this bomber concept was allowed guarantees that it bears little resemblance to the planned Stealth Bomber. Concealed air intakes and engine exhausts, plus the absence of vertical surfaces, reduce the radar target area.

Below: Modestly concealing its modified "stealthy" air intakes from the camera, the B-1B displays its clean lines. Requests from a well-known aviation magazine for details of the new intakes were politely refused by the USAF!

Above: When developing the SR-71, and the A-12 from which it was derived, Lockheed engineers devised a radar-absorbent material able to withstand the thermal stresses of Mach 3 flight. It is used on the wing leading edges and elevons.

posites are being used in US stealth projects, and the importance of these materials may be judged by the tight-lipped response to any enquiries relating to the electrical properties of advanced composites.

Used on their own, such materials will not create "invisible aircraft", since any aircraft will still contain a fair proportion of metal parts: the British de Havilland Mosquito light bomber of World War II had a wooden fuselage, but could still be tracked by German radar. Further work is needed in order to create stealth aircraft and missiles.

Contour control

Careful control of aircraft shape allows designers to avoid flat surfaces and sharp angles – features which serve as good reflectors of radar energy. Smooth and blended shapes are used with the aim of deflecting rather than reflecting the incoming radar energy. Some evidence of this technique may be seen in the shape of the Rockwell B-1 bomber. The slab sides and relatively straight lines of the B-52 gave that bomber a high radar cross section, but the rounded lines and wing-body blending of the B-1A, taken even

Aircraft inlet configurations

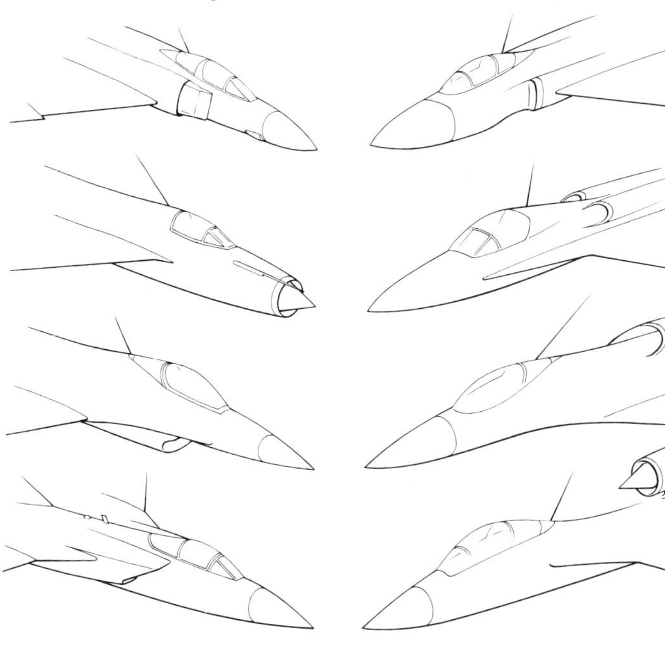

Above: Current patterns of fighter use patterns of air intake which can act as good reflectors of radar energy. Shown on the left are (from top) the F-4, MiG-21, F-16 and Tornado. Stealth aircraft are likely to sacrifice intake efficiency in favour of reduced radar cross section. Potential configurations shown on the right include flush intakes (top), and various patterns of dorsal inlet which the fuselage will screen from the attention of ground-based surveillance and tracking radars.

further on the B-1B, are designed to minimize radar reflections.

Since flat surfaces act as good radar reflectors, stealth designs tend to eliminate features such as the vertical stabilizer, or at least to tilt them away from the vertical. Winglets or butterfly tails have been suggested as alternatives to conventional fin and stabilizer arrangements. Even the wing has come under scrutiny, with tests being carried out on high-mounted wings which may be turned through 90° once the aircraft reaches cruising speed, then stowed conformally on the upper surface of the fuselage. Body lift would sustain the aircraft in high-speed flight.

Not all components of an aircraft are amenable to large-scale reshaping. Wing and tail leading edges, air intakes and other structural features can act as highly efficient reflectors of radar energy. Intakes may be designed to use half-cone centre-bodies, plus kinked or even zig-zag air trunking, techniques which prevent radars from "seeing" the highly reflective front face of the jet engines, while well-rounded leading edges may be substituted for the sharper-fronted planforms normally used, but residual reflections are bound to remain.

Radar-absorbent materials

Such residual reflections may be reduced by means of radar-absorbent materials – the electronic equivalents of the black finish used on the undersides of World War II night bombers. In order to cope with

Aircraft target signatures

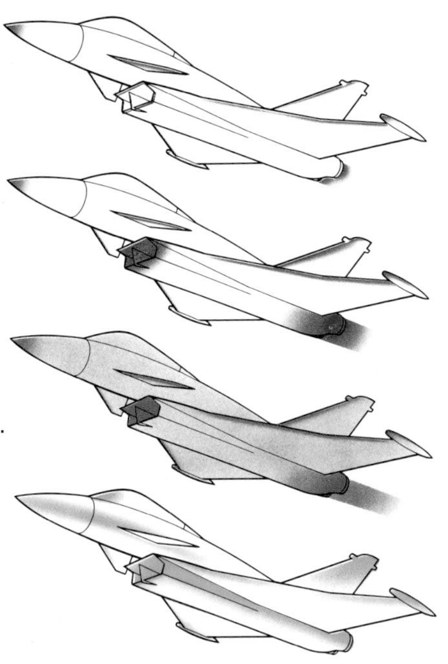

Right: The target signature of the proposed European Fighter Aircraft is both complex and classified. Our prediction is that at 2-3 micron IR wavelengths, energy will be emitted only from the hottest parts of the target — nose and jetpipes at subsonic cruise (top), while with the use of afterbuner the hot tail area will grow in size, and the inlet system will become an attractive target. At longer IR wavelengths such as 10.6 microns most of the airframe will be emitting energy. Radar signature is more difficult to predict, but strongest returns will probably come from the areas shown in blue.

the wide range of frequencies used for radar, broadband absorbent material must be used. The types most commonly used in the laboratory are effective, but bulky. Plessey, for example, manufactures low-density foam material impregnated with carbon-loaded plastic. This allows the radar signal to enter the foam without reflection, before being absorbed after only a short travel within it.

For operational applications more compact materials are required. Plessey's LA1 flat-sheet absorber is around 0.5in (12mm) thick, and absorbs more than 97 per cent of the 5.5-30GHz energy which strikes it. The company has also devised a radar-absorbing camouflage net which can reduce radar signatures by a factor of more than 8 at 6-100GHz, and by 30 or more at the widely-used I/J band frequencies.

Radar-absorbent material has in the past been relatively heavy, so could not easily be applied to large areas of an aircraft, and its use was limited to critical parts which had an unduly high radar reflection. Recent developments are reducing this weight penalty however — the LA1 sheet mentioned above weighs only 0.6lb/sq ft (0.3kg/m²).

RCS reduction

Application of all these techniques dramatically reduces the radar cross section (RCS) of even large aircraft such as bombers. When approaching head-on to a hostile radar, the B-52 has an RCS of more than 1,080sq ft (100m²) while the more rounded B-1A offers a target when seen head-on of only 108sq ft (10m²). Further refinement of the basic design such as removal of the dorsal spine formerly planned as a housing for ECM antennas and equipment, reshaping the leading edges of the flight-control surfaces and modifying the engine inlets has reduced the cross section of the B1-B to around 11sq ft (1m²) or less.

A typical jet fighter such as the F-4 has a radar cross section of around 21sq ft (2m²), but in the new stealth designs this is reduced by a factor of 100 or more to 0.2sq ft (0.02m²) or less. Even this figure may be open to

Discernible emissions

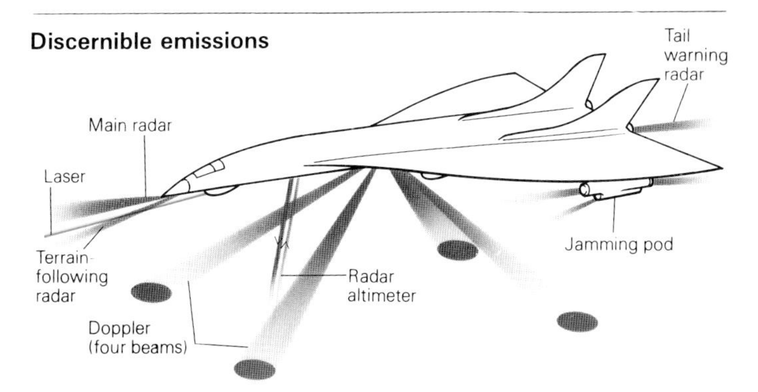

Above: If a stealth aircraft is to avoid detection by enemy ESM systems, it must avoid emitting the signals shown above.

improvement. The US magazine *Defense Electronics* quotes an engineer involved in stealth bomber development as saying, "We're seeing radar cross sections of less than one millionth of a square metre".

These figures will rise when external stores are carried, since a 500lb (230kg) iron bomb or an AGM-65 Maverick air-to-surface missile probably has a radar cross-section of 1sq ft ($0.1m^2$) or more. Stealth aircraft will have to carry ordnance within internal weapon bays or in semi-buried conformal locations covered with radar-absorbing fairings.

Thermal signature reduction

With the growing use of IR sensors as a radar substitute, measures must be taken to reduce the thermal signature of a stealth aircraft. Performance is likely to be subsonic, since the levels of airframe heating induced by supersonic flight would make the aircraft an easy target for IR sensors. The powerplant poses further problems, since its hot aft section must be screened from view, and the hot efflux from the engine core has to be mixed with cooler air.

One propulsion system which greatly reduces IR emission involves ducting the engine efflux along the wing leading edge, then ejecting it over the top surface of the wing via a slot nozzle extending along some three-quarters of the wing span. This has been tested in the USA using the Ball-Bartoe Jetwing single-seat research aircraft. Powered by a single JT15D-1 turbofan, this features a small wing panel mounted above the slot nozzle. The air passage between the main wing and this smaller upper surface acts as an ejector, providing thrust augmentation and mixing the engine efflux with ambient air, reducing IR signature.

Another technique under study is the jet flap, which involves releasing the engine efflux at the trailing edge of the wing flaps via a nozzle with a high length-to-height ratio. For a fighter design, this would incorporate a two-dimensional afterburner. According to US studies, the resulting aircraft would have a low IR signature, plus a reduced rear aspect radar cross section.

The prospect of laser-based radars operating at infra-red or even visible light wavelengths could threaten current stealth aircraft, but suitable technology to handle such systems is already emerging from the laboratory. Absorbent materials have been developed for use at infra-red and sub-millimetre wavelengths, the result of research into baffles able to trap stray radiation within sensors operating in this area of the spectrum. One multi-layer treatment consisting of a polyurethane binder and carbon-pigmented multi-layer coating has demonstrated a reflectance of less than 10 per cent.

Emission suppression

The techniques described so far are purely passive, being intended to reduce the radar and infra-red

Above: This Grumman concept is intended for high-altitude reconnaissance, so is designed for minimum signature when seen from below. The flat lower surface is intended to reduce radar cross section, while the inlet and engine exhausts (prominent radar and IR targets) are shielded by the fuselage.

Hypothetical stealth STOVL fighter

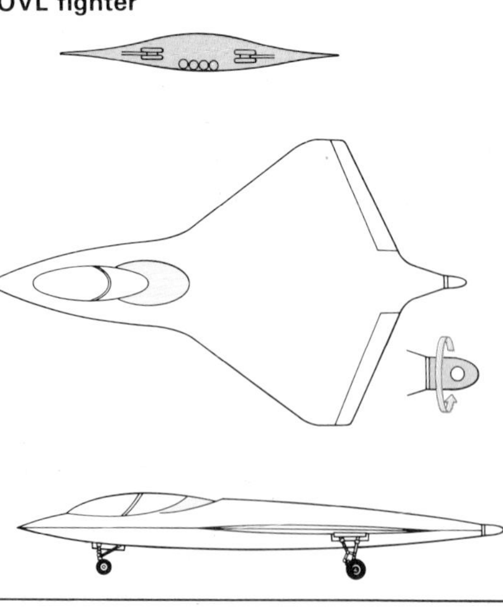

Right: Aviation author Bill Gunston has devised this hypothetical STOVL fighter configuration to illustrate the use of stealth technology. The flush-mounted intake just aft of the canopy, absence of vertical surfaces and the use of wing/body blending all minimize radar echoing area. The armament would be carried internally to avoid spoiling the aircraft's anti-radar qualities, while a tail-mounted rotating reaction control nozzle would give high agility.

signature of the aircraft, but equal effort must be made to reduce the detectability of all radio and radar transmissions made during a mission. Conventional radio and radar transmissions are prime targets for ESM sensors, so stealth aircraft will either use passive sensors which do not involve transmission, or will rely on spread-spectrum techniques to "bury" the transmission among the normal background of radio and radar signals making them difficult to detect and identify. A joint USAF/DARPA project to develop a stealth fire-control radar for airborne applications is reported to use both millimetre-wave and conventional I/J-band frequencies.

For short-range air-to-air communications, stealth aircraft might use low-power millimetre-wave transmissions at frequencies which are rapidly absorbed by water vapour within the earth's atmosphere. Highly-directional but compact "smart" antennas could be used to beam the radio energy accurately in the direction of the intended recipient, while the narrow beamwidth and high attenuation would make it difficult for ESM systems not directly on the line-of-sight of the transmission to detect and intercept the signal.

Specialized ECM

Stealth aircraft will also rely on the use of electronic countermeasures. The EW techniques described in this book are equally applicable to the self-defence of stealth aircraft, but since stealth aircraft return only a minute radar echo, the power levels needed to mask this are much smaller than for conventional aircraft. SInce low-powered jamming transmitters will take up less of the avionics space and weight allocated to EW systems, a stealth aircraft will be able to allocate more of its EW payload to signal-processing circuitry, allowing the use of highly-sophisticated deception techniques in order to mislead hostile radars as to the size, nature, position and course of the aircraft.

USAF stealth aircraft

Several patterns of stealth aircraft have been test flown by the USAF, mostly from a classified site in Nevada. These are thought to be technology demonstrators rather than prototype fighters and to be in the 20,000lb (10,000kg) class. Since flights started in 1979, at least two have crashed. Designations reported for stealth projects include F-19, at least one YO-series number, and CSIRS (Covert Survivable In-weather

Hypothetical stealth bomber

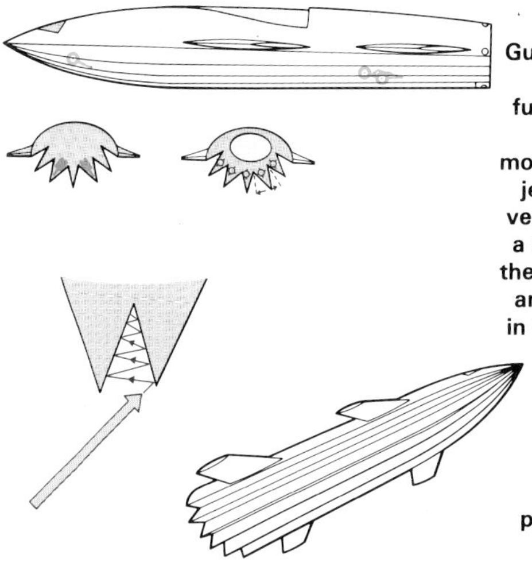

Left: Another Bill Gunston design explores far-out concepts for a future bomber. Vestigial wings, the use of aft-mounted reaction-control jets, and an absence of vertical surfaces or even a canopy, together with the stowage of ordnance and undercarriage within wedge-shaped ventral strakes, all minimize radar echoing area. Radar energy striking the belly would be trapped between the strakes, losing 30-40 per cent of its strength at each reflection.

Reconnaissance Aircraft). The latter may be intended for operational use.

Lockheed F-19

Best-known US stealth aircraft is the Lockheed F-19. Originally known as the XST (Experimental Stealth Tactical), the Lockheed stealth fighter first flew in 1977, and probably took part in a competitive fly-off against rival designs. Ordered into production in 1981 as the F-19, the aircraft may be designated Aurora – a name which could easily be explained away to anyone catching an unauthorized glimpse of F-19 documentation, since Lockheed has supplied CP-140 Aurora maritime-patrol aircraft to the Canadian Armed Forces.

The first F-19 squadron probably formed in 1983/84 at the Groom Lake facility at Nellis AFB, California. Few outside eyes have glimpsed the new fighter, whose basic configuration –

The shape of tomorrow's aircraft?

Right: This Bill Gunston "F-19" combines many of the features discussed earlier (such as flush intakes and internal armament) with the few details of the new fighter which have emerged to date. It should be taken as an impression of the general concept rather than as a prediction.

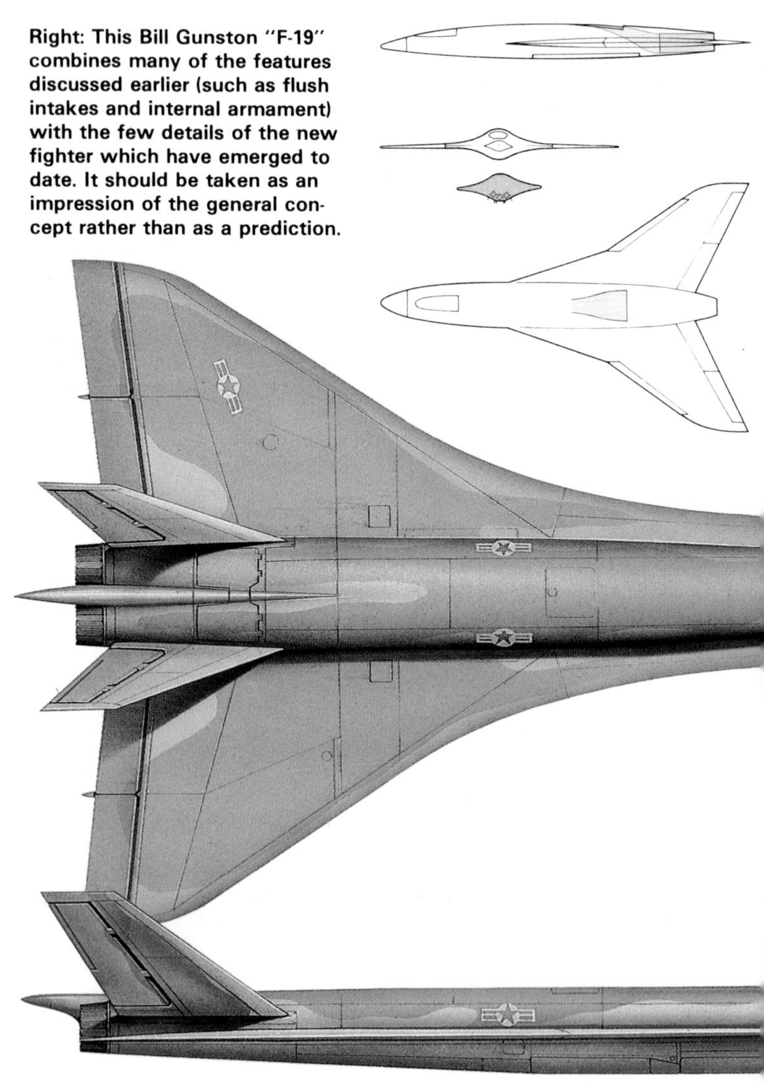

a large and rounded fuselage blended with a relatively small wing – has been liked to that of the Space Shuttle or even of the NASA lifting bodies tested in the late 1970s.

Northrop ATB

Much less is known concerning the Northrop Advanced Technology Bomber. The company will only admit that it has been given a contract to perform initial research and development on "advanced bomber concepts". Further details of the programme are highly classified, but some clues may be gleaned from known areas of Northrop expertise.

Many experts predict that the ATB will be of flying-wing configuration, an idea which may be supported (or might even have originated) from the fact that shortly after award of the USAF contract, Northrop vice-president for advanced projects Lee Begin wrote a paper on the history of the unsuccessful Northrop YB-49 Flying WIng of the late 1940s. Northrop also has significant experience in carbon-fibre technology and is a major supplier of ECM systems.

ATB is expected to be smaller than the B-1, and will probably have a metal load-bearing structure. First flight is planned for 1987, and production of 110 or even up to 150 ATBs is expected to follow close behind that of the B-1B, with the aircraft entering operational service – presumably as the B-2 – in 1992. The Northrop bomber will replace the B-1B in the penetration role, maintaining SAC's ability to penetrate Soviet air defences into the 21st century.

Soviet response

Soviet designers are unlikely to surrender meekly to the threat posed by US stealth aircraft. In addition to starting its own stealth programme, the Soviet Union is likely to develop alternative methods of tracking targets. Even the atmospheric turbulence created by the aircraft's passage through the air might be utilized as a trackable "signature".

Some experts even doubt the long-term future of stealth aircraft. Dr Edward Teller, "Father of the US H-bomb", has expressed the opinion that "The American people are being misled about the possibility of a practical and effectively invisible bomber". Writing in *The Wall Street Journal,* he warned that "Countermeasures seem so easy that even a couple of years – considerably less time than the actual production time requirement for stealth – would suffice for their deployment".

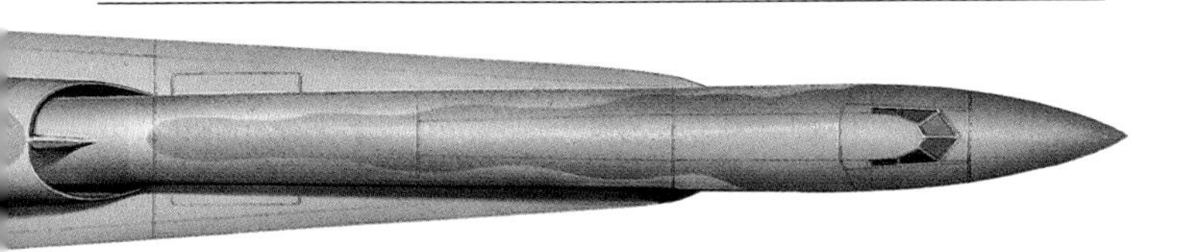

Above and below: This Bill Gunston concept suggests one possible approach which Northrop may be taking to the Advanced Technology Bomber project – minimal wing area, dorsal intakes, and angled vertical tail surfaces. Already used by the SR-71, the last-mentioned are poorer radar targets than conventional tails. The author of this book believes that a delta-wing planform could be adopted for the ATB, but considers such a prominent fuselage an unlikely feature. A minimal fuselage heavily blended into the wing is more likely.

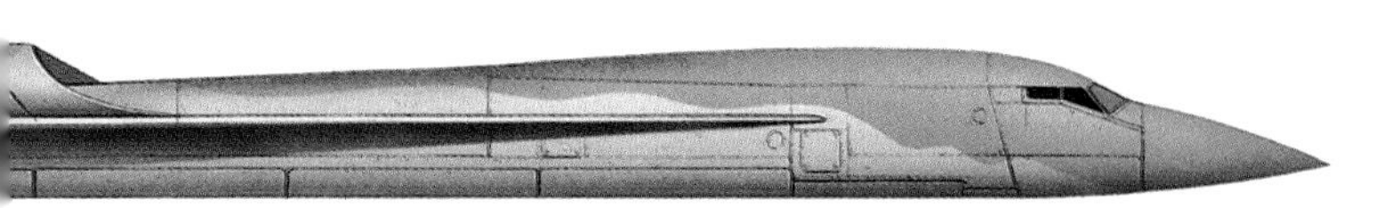

Electronic Intelligence Gathering

SIGINT (signals intelligence), elint (electronic intelligence) and comint (communications intelligence) are activities which many nations practise, but to which few will admit. Researching a recently published book on EW, one author was assured by the Royal Air Force that the UK does not operate elint aircraft – a claim that might surprise the Royal Swedish Air Force, which several years ago released a photo of an RAF Nimrod R.1 caught practising its electronic art near Swedish airspace. Externally similar to the normal Nimrod ASW aircraft, the R.1 lacks the former's elongated tailcone: used to house the magnetic anomaly detector (MAD), this is not required in an aircraft which never attempts to locate underwater targets.

Sigint is a passive activity, a process of monitoring and analysing intercepted electromagnetic signals. It is normally divided into two types of activity. Comint covers the interception, analysis and decryption of radio communications, while elint involves the study of radar and other non-communications signals. These terms are often used loosely by non-sigint personnel, and many publications use the term "elint" to cover all forms of sigint activity, referring for example to all sigint aircraft as "elint aircraft".

Tactical and strategic sigint

Military sigint operations may be either tactical or strategic. Tactical sigint is intended to locate and identify weapon systems and military units of a potential or actual enemy, while strategic systems monitor the overall equipment and deployment of an opponent's forces, and may give timely warning of the development of new equipments, technology or tactics.

Targets for sigint monitoring include communications and air-traffic control, aircraft, spacecraft, surface vessels, submarines, command centres and ground forces. By careful monitoring of signals it is possible to establish the types of radar, fire-control systems, communications systems and other threats which an opponent has fielded, or may be about to deploy. In the latter case, the research and test facilities of potential enemies will have to be

Above: A Soviet AGI keeps a dangerously close watch on the US carrier *Franklin D. Roosevelt* in the Mediterranean.

targeted for elint coverage, and such establishments may be located far from the physical border, disguised, and consequently difficult to keep under observation.

Other features worthy of note might be new operating frequencies and new antenna-scan or modulation patterns – all may be evidence of new equipments or upgrades to existing types. Analysis of the parameters of signals from threat radars will provide the data needed by designers of deception jamming equipment.

Details of enemy strategy and tactics may be gleaned by observing the location, density and movements of hostile emitters. Their location and density can serve as indicators of deployment patterns, while high concentrations of signals in remote areas of a country may disclose the existence of a new base, or provide first warning of highly classified enemy operations. Knowledge of the location of radio and radar emitters associated with major weapon systems allows the compilation of an Electronic Order of Battle.

Sigint organizations

In 1947 the governments of the USA, UK, Australia, Canada and New

Below left: USAF elint U-2Rs carry mission equipment in the wing-mounted "superpods".

Below: CRT display and keyboard of the Thomson-CSF Elisa elint receiver. The operator is analyzing an intercepted signal.

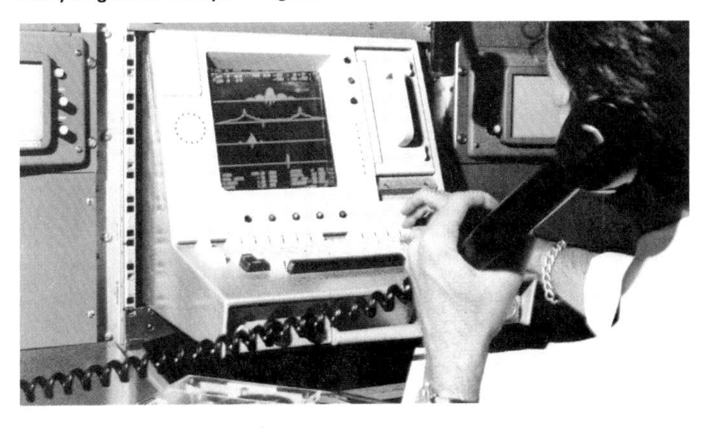

Above: Britain's Government Communications Headquarters (GCHQ) at Cheltenham handles UK sigint tasks.

Above right: Panoramic display of an EM Systems R400 receiver, showing five frequency bands.

Right: Despite computer-aided analysis, operators must still listen to intercepted signals.

Zealand joined forces under the UKUSA Agreement to divide between themselves the task of maintaining sigint coverage of the world. Pooling resources in this way improved coverage and reduced duplication, with the resulting data being shared. Most of the work was shared by the two largest Western sigint organizations – the US National Security Agency (NSA), and Britain's Government Communications Headquarters (GCHQ).

Headquarters of the US National Security Agency is a vast complex at Fort Meade, Maryland, between Washington D.C., and Baltimore. This houses facilities such as analysts and their complex computer systems, including the Cray-1 supercomputer, an operations control centre, a secure printing plant for highly classified material and storage facilities for sensitive matter. According to the General Accounting Office of the US Congress, the NSA produces more classified material than all other branches of the US Government combined – including the armed forces, the State Department and the CIA.

Britain's GCHQ is based at Cheltenham, Gloucestershire, but maintains intercept stations at around a dozen locations in the UK. Identified locations include Beaumanor, Brora, Cheadle, Flowerdown, Gilnahick, Hawklaw, Irton Moor, Knockholt, Sandridge and Shaftesbury.

After its formation in 1952, the NSA moved quickly to establish a worldwide network of listening sites. By the mid-1950s it was working to a plan which called for the installation of more than 4,100 listening posts. The US Army, Navy and Air Force also maintain sigint stations, many of which existed long before the NSA was created.

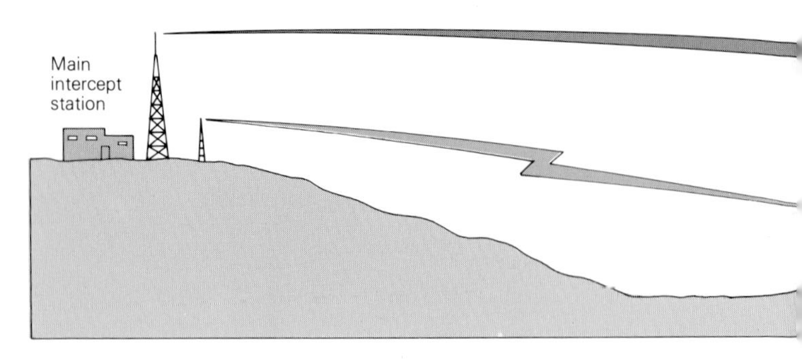

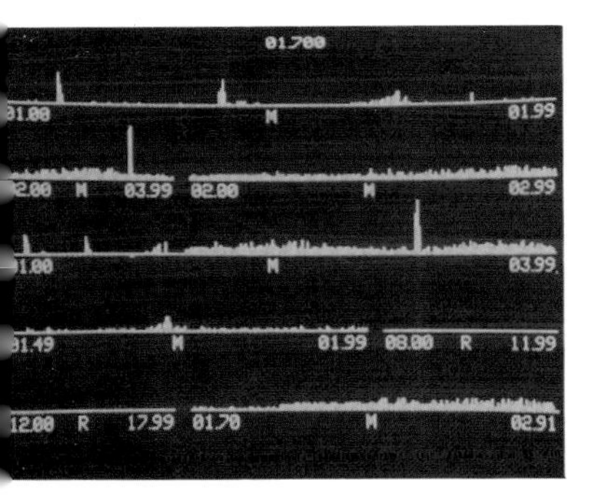

Comint

Most armies invest considerable efforts in the area of comint – the passive monitoring of enemy transmissions. Interception systems are used to detect transmissions, a process normally followed by real-time analysis. Once a communications net has been detected by ESM operators, a number of tactics may be employed.

The brute force solution is to use jammers to disrupt the network, denying the enemy units their normal communications. A more subtle approach is to leave the net to operate normally, so that monitoring of the signals can provide intelligence information or tactical surprise.

Skilful Egyptian comint during the Middle East War in 1973 attempted to foil the retreat of Israeli personnel from a besieged position on the Bar Lev line. After marching for an hour under cover of darkness, soldiers retreating from one position transmitted a codeword by radio, then prepared to fire a green flare to signal their position to an AFV tasked with meeting them in the desert – tactics agreed after an earlier radio conference with higher command which the Egyptians had overheard. Up went the signal flare – but from Egyptian lines!

HF and VHF interception

Below: Thanks to the long range of HF radio signals, land-based intercept and direction-finding stations (left) may be targeted on transmitters deep in hostile territory, while semi-fixed sites closer to the border intercept the shorter-ranged VHF signals.

Intercepts
Target VHF communications
Target HF communications
EW tasking/feedback links

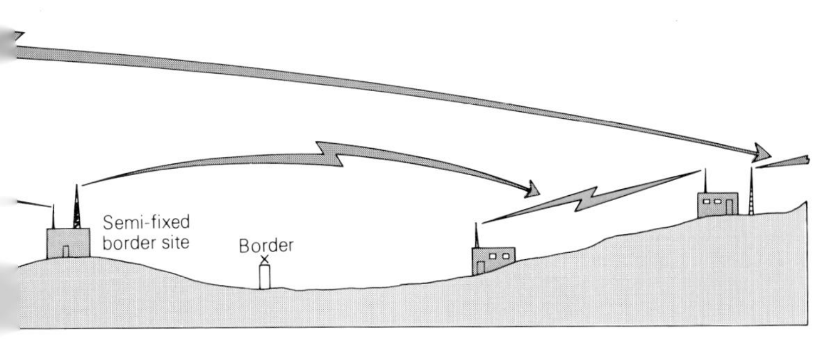

When the commander of the retreating unit realized what was happening, he warned his HQ of the deception, then fired two flares to mark his true position. Given this genuine fix the advancing tank was able to change direction and make a successful rendezvous.

Comint targets

Diplomatic traffic is considered a legitimate target for comint, but with the growing use of sophisticated cryptographic equipment, decoding of transmissions is becoming ever more difficult. Soviet high-level cryptographic systems are for all practical purposes unbreakable. Even the statistical analysis of traffic patterns and transmission conventions (a technique which can produce some useful data from unbroken intercepts) yields little, since Soviet operating practice on high-level links is to maintain the flow of data at all times, effectively burying the genuine data in an endless flood of enciphered random characters. The US Government uses equally secure systems and operating methods.

Most developed nations are probably difficult comint targets, given the widespread use of advanced cryptographic equipment, so the NSA and GCHQ probably have more success in reading Third World signals. Britain is reported to have monitored Argentinian military and

diplomatic communications during the Falklands War. A certain amount of sigint work is even targeted against allies – the US Army Security Agency has apparently intercepted communications from the British Embassy in Washington.

A significant portion of comint work may be against industrial targets, since both the NSA and GCHQ monitor telex communications from their respective countries, and can thus read the commercial secrets of private companies.

Below left: The Emerson MSQ-103A Teampack Dragoon monitors from 500MHz to 50GHz.

Below: The British Army's C50 UHF radio relay is mobile, highly directional and hard to jam.

Speech recognition

The sheer amount of radio traffic, particularly of low-grade routine communications, is vast – detailed assessment would strain the monitoring and translation facilities available even to the superpowers. One possible solution would be to use computerized speech-recognition equipment to scan tape recordings of intercepted communications in an attempt to find predefined keywords.

Although this may well be beyond the ability of current-generation

systems, some observers already credit the UK security forces with having developed and deployed such equipment for purposes such as telephone tapping. It may be significicant that the Logica Logos speech recognition system now being incorporated into some items of British military electronics was originally developed for the Joint Speech Recognition Unit – a little-known establishment closely connected with GCHQ.

Radiation intelligence

One rarely discussed technique is rint (radiation intelligence), the gathering of intelligence by monitoring stray, non-information-carrying radiation from equipments or systems. Vehicle ignition systems and even electricity power lines have been mentioned as possible sources of rint data.

One now-declassified EW textbook suggests that rint equipment may locate a radar which has temporarily shut off its transmitter to avoid being detected by elint sensors. This might be done by detecting radiation from the radar's electrical generator. If the set has not been shut down but is transmitting into a dummy load (an antenna simulator which emits no signal), radiation might even be detected from the pulse-handling circuitry within the transmitter. Modern design techniques and anti-interference legislation attempt to limit such stray radiation, but rint may still play a useful role, albeit probably of limited value, in contemporary military operations.

Above: Circularly disposed antennas arrays, such as this US site, can direction-find at high, very high and ultra high frequencies with an accuracy of between 3° and 5°.

Right: This 1975 SR-71 reconnaissance photograph shows the rhombic antennas of an HF communications site at Berbera in Somalia.

Geopolitical interference

The US sigint network was designed to ensure global coverage, but one effect of the unrest in the Middle East since the early 1970s has been to disturb some of the most sensitive facilities intended to monitor Soviet missile trials and space shots.

The original US sigint site at Karamursel, in Turkey, was supplemented by smaller sites around the country. Among the most important were Trabzon, only 70 miles (113km) from the Soviet border, Samsun, used to monitor the Soviet missile test facility at Kapustin Yar, and Diyarbakir, which covered the Tyuratam missile range. These facilities were temporarily denied to the US in retaliation for an arms embargo placed on Turkey following the 1974 invasion of Cyprus, staying closed from 1975 until 1978.

This hiatus in sigint coverage was no sooner solved, when the Iranian

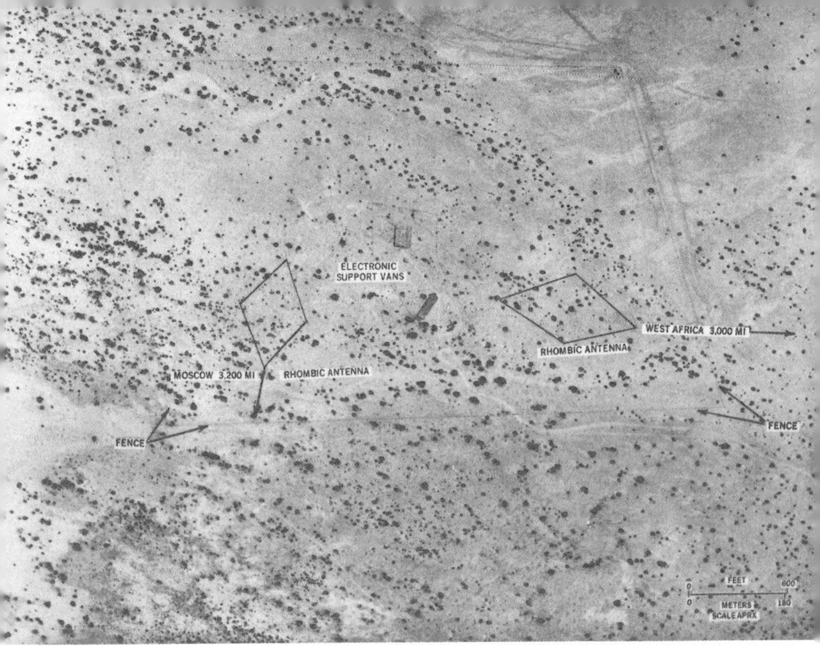

revolution in 1979 resulted in the US losing even more valuable monitoring stations in Iran. Located close to the coast of the Caspian Sea, these were invaluable monitors of Soviet missile trials. By 1980 a new sigint agreement — this time with the Chinese Government — had seen the construction of a new listening station in Western China.

Antennas

Sigint antennas for use at HF and below are large in size. Omnidirectional antennas include the Wullenweber type, whose four concentric circular arrays of masts and antenna wires have a maximum diameter of almost 9,000ft (2,750m). The outer ring is the high-band array, while the next ring acts as a reflector for the outer, increasing its sensitivity from signals approaching from the front, and reducing the effect of signals coming from the rear. The two rings nearest the centre form another receiving array and associated reflector, this time for the lower frequencies.

Since the receiving rings are made of several segments, measurement of the phase of the signals received by each segment (effectively the tiny differences in time between the signal's reception by the various segments) allows the electronics to determine the bearing of the signal source. Bearings from two or more geographically separated stations allow the position of the emitter to be determined.

Directional antennas such as the Rhombic have been used for many years for radio communications. These are sensitive only to signals from a specific direction, so are ideal for communications between fixed sites, being laid out at the time of construction in an orientation which will aim its directional qualities in the direction of the site it will communicate with. For sigint purposes, such arrays must be custom-built so that they are aligned with the target of interest.

At higher frequencies, antennas are much more compact, ranging from Yagi arrays used at VHF and UHF (the same type as that used for domestic TV), to blade antennas, conventional dishes and flat planar arrays. They are thus easy to mount on vehicles, ships, aircraft or satellites.

Signal analysis

Receivers designed for elint purposes (usually known as "ferret" receivers) must incorporate a number of features not found in most general-purpose units. Frequency coverage should be as wide as poss-

Above: The deck houses on the Primoye class AGI *Zakarpatye* are probably for sigint equipment.

ible, while signals using all possible forms of modulation should be detected and recorded, and measurements must be carried out accurately if the resulting data is to be used for the design of ECM systems. Basic measurements made by elint receivers include:

- strength of the received signal
- fluctuations in signal strength due to factors such as antenna scan pattern or on/off switching
- polarization of the signal
- modulation pattern (the information superimposed on the signal)
- bearing of the signal.

Each radar or other emitter has its own characteristic "signature" —

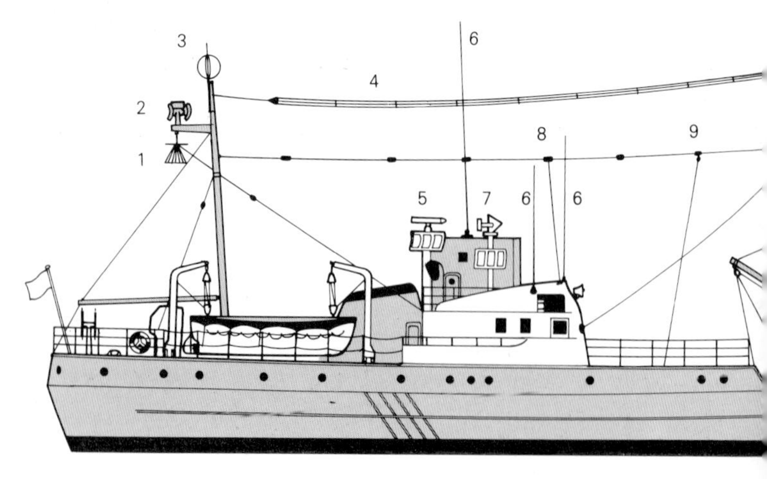

effectively an electronic "fingerprint" by which it may be recognized. NATO reporting names for many early patterns of Soviet radar were based on the distinctive sound which the intercepted signal made in the elint operator's headphones. "Owl Screech", "Fan Song", and "Scan Odd" are typical examples of such designations.

Signal densities are normally so high that the target signal may be observed against a background of other transmissions. In most cases the elint operator must use his skill to sift through the signals being detected, deciding in real time which should be recorded for later analysis. While carrying out this work, elint operators must always bear in mind the possibility that both they and their equipment may be attacked by a weapon guided by some of the signals they are receiving.

Antenna deployment

The sheer size of the antennas used to monitor signals at frequencies of up to and including HF band demands large amounts of real estate, so equipment of this type can only be installed at fixed sites. Since the radio signals being intercepted have very long range, a small number of stations can cover large areas of the globe.

For sigint at higher frequencies, where transmissions are not reflected from the earth's ionosphere, the receivers must be much closer to their targets. Mobile platforms such as ships, aircraft, or even satellites are required, usually relying on sideways-looking antennas for the observation of targets of opportunity from stand-off range. Aircraft may also be used to overfly areas of interest, but this is normally possible only in wartime. The only vehicle which may carry out overflights with impunity is the orbiting satellite.

The Soviet elint fleet consists of around 40 unarmed vessels based on commercial fishing trawler designs. These are familiar sights during Western naval exercises, and are often stationed offshore near specialized installations such as Cape Canaveral.

US Navy sigint ships

An old Chinese curse says "May you live in interesting times", and this turned out to be the fate of the US Navy's sigint ships, built as wartime transports and pressed into service as unarmed listening posts. First-generation vessels *Valdez*, *Muller* and *Robinson* were taken into service in the early 1960s in order to improve sigint coverage of Africa and South America. Crewed by civilians of the US Navy's Military Sea Transportation Service, they were soon supplemented by the USN-manned *Oxford*, *Georgetown*, *Jamestown*, *Belmont* and *Liberty*.

The first casualty was *Liberty*. Sta-

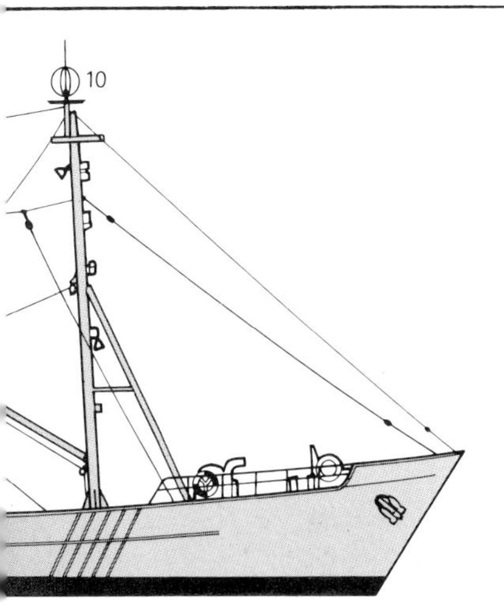

Soviet Okean class AGI (elint trawler) communications fit

More than 100 Okean class trawlers were built in East Germany between 1959 and the mid-1960s. Only *Linza* and *Zond* have the particular fit shown.

1 "Disc Cone" omnidirectional receiver antenna
2 Radar receiver
3 Direction-finding loop
4 Folded dipole for HF reception
5 "Don 2" navigation radar
6 Vertical rod antennas
7 Radar receiver
8 Coaxial feeder to wire dipole antenna
9 Wire antenna apron for dipoles, end feeds, etc.
10 Direction-finding loop with earthplane

tioned off the Egyptian coast during the 1967 Six-Day war, she was attacked and badly damaged by Israeli fighters and torpedo boats less than 24 hours before the launching of Israel's counterattack on the Golan Heights. Israel eventually paid compensation for the damage to the vessel and the deaths of 32 of its crewmen, but full details of the incident – and the reason for the Israeli attack – remain classified by both governments.

The USN added shelter-mounted sigint electronics to standard destroyers and destroyer-escorts in order to create further sigint vessels, two of which may have featured in the opening incidents of the Vietnam War. The destroyers *Maddox* and *Turner* were apparently on sigint duty when allegedly attacked by North Vietnamese torpedo boats.

The next generation of sigint ships comprised the *Banner*, *Pueblo* and *Palm Beach*, but these had a short life. Following the North Korean seizure of *Pueblo* on January 5, 1968, the US Navy sigint fleet was phased out. Stripped of their specialized electronics, the vessels were sold for scrap.

Although the era of dedicated US Navy sigint ships is over, the service's warships can still maintain a watch on targets of interest. These operations may involve add-on sigint systems, but it is more likely that the growing complexity of naval ESM systems gives equipment of the latter type most of the capability demanded by sigint operations.

Airborne platforms

High-flying aircraft make good sigint platforms. US types used for this role include the Lockheed EC-130 variant of the Hercules transport, the Lockheed U-2/TR-1 series, the Lockheed SR-71 Blackbird, and various models of Boeing C-135 (including EC-135, RC-135B, -135C, -135D, -135E, -135M, -135S, -135U, and -135W), plus the EP-3B and -3E variants of the US Navy's Lockheed P-3 Orion. Other long-range US aircraft suspected of having a sigint role include the Boeing RC-135T and -135R, plus the Lockheed C-130A-II and C-130B-II.

In addition to these aircraft, NATO has further airborne sigint platforms in the shape of the Italian Air Force's Aeritalia G.222VS, several Hawker Siddeley Nimrod R.1 "radio calibration aircraft" of the Royal Air Force and five modified Breguet Atlantics operated by West Germany.

Tactical sigint aircraft such as the US Army's Beechcraft RU-21 series and the Israel Aircraft Industries IAI-201 Arava operate much closer to the emitter being observed, monitoring or even jamming communica-

Left: The damaged sigint ship *Liberty* arrives at Malta eight days after being attacked by Israel during the 1967 War.

Below: The North Korean seizure of the *Pueblo* in January 1968 exposed weaknesses in US command and control systems.

Top: "Cub-B" elint variant of the An-12 transport carries a large number of antennas but only civil registration.

Above: When this 1971 photo was taken, this RC-135U still showed some -135C features.

tions. During the Vietnam War, Douglas EC-47 sigint aircraft operating under the Combat Cougar programme located enemy ground units which were subsequently attacked by fighter-bombers or even B-52s. Helicopters may also be used as short-range sigint platforms. The USAF operates the Bell EH-1H Iroquois and plans to operate the EH-60 version of Black Hawk.

Drones and RPVs are also suitable for sigint duties, the USAF having operated types such as the Ryan 147 recce/sigint drone over China during the early 1960s. During the Vietnam War, drones flew more than 3,000 missions, although not all were for elint purposes. For the latter task, Teledyne added wing tanks for extra fuel and facilities for data telemetry to the basic Model 147 RPV, creating the BQM-34R high-altitude model. This is reported to have flown missions of six hours or more. During the Compass Dwell and Combat Dawn programmes, RPVs were used to locate the positions and operating frequencies of North Vietnamese radar sites.

Soviet sigint aircraft range from the obsolescent piston-engined Il-14, through "Cub-B" and "-C" versions of the Antonov An-12, the "Coot-A" version of the Il-18 airliner, "Badger-F" and "-K" versions of the Tu-16 bomber, and "Bear-C/D" multipurpose versions of the Tu-142.

The art of provocation

Many successful sigint operations involve a degree of provocation. If useful data is to be obtained, the victim must be persuaded to activate his defences, resorting to operational frequencies and tactics. Aircraft can, for example, be despatched on flight paths directly heading towards the airspace of the target nation, in order to resemble a potential intruder. Such ferret flights sometimes end in disaster — the victim may decide to shoot down the aircraft, either because it was assessed as a genuine threat or because the victim wished to dissuade the operator of the aircraft from attempting future missions of this type.

The Soviet Union is well known for its "shoot first" policy towards ferret aircraft, or any other intruders into its airspace — between 1950 and 1969, the Soviet air defences opened fire on 23 US aircraft, downing 12. Operating close to or in some cases within Soviet airspace, most were probably on ferret missions.

How effective such airborne elint platforms can be was illustrated around 1950, when a British elint aircraft ended months of useless speculation on whether or not the Soviet Union had operational air-intercept radar by making a 20sec recording of signals from a "Scan Odd" radar, presumably the result of a provocative elint mission which had risked interception in order to sample the interceptor's radar transmissions.

Bluff and counter-bluff

Such sigint probing remains a struggle between bluff and counter-bluff. Faced with simulated threat, the victim may choose to accept what amounts at worst to a low risk of attack, and may opt not to activate his defences, or might even switch to frequencies or operating modes which do not correspond with any planned wartime procedures.

Several US military aircraft losses during the Cold War years may have been a result of such provocative missions. One widely respected defence journal went so far as to suggest that the Korean Airlines Boeing 747 shot down by a Soviet Air Force Su-15 "Flagon" in 1984 had deliberately flown into Soviet airspace as a part of a Korean/US sigint operation. The proffered evidence was untenable, and must be dismissed as an exercise in coincidence-hunting. Having been taken to court by Korean Airlines, the journal subsequently withdrew its allegation. There is no evidence that commercial airliners have ever been used in provocative sigint operations against the

Above: Operators monitor their electronic equipment consoles aboard a USAF NKC-135.

Soviet Union, although the US has on several occasions claimed that Aeroflot aircraft have overflown sensitive Western installations.

Lunar reflector

The earliest attempts to monitor deep within the Soviet Union used manned aircraft, but the development of surface-to-air missiles promised to end the usefulness of this technique. The obvious alternative was the use of satellites, although one bizarre scheme started in the late 1950s involved attempts to construct a 600ft (183m) radio telescope at Sugar Grove in West Virginia. Officially for radio astronomy, this would have attempted to use the moon as a radar reflector, detecting the tiny amount of energy from Soviet radars and high-powered radio communications systems. The technical problems involved in constructing a radio telescope more than twice the size of Britain's contemporary 250ft (76m) Jodrell Bank instrument proved insuperable and the project was abandoned.

The Sugar Grove site remains in operation, however, and is operated by the US Naval Security Group. Now equipped with several smaller "dishes", its role is probably to monitor the flow of data between the Communications Satellite Corporation's spacecraft and its ground station at Etam in West Virginia.

Intelligence-gathering in Space

All US intelligence-gathering spacecraft are classified, but analysis of their orbits and behaviour, plus leakages of classified information and a series of spy scandals, have brought some details into the open. Some experimental sigint equipment may have flown in 1960 or 1961 aboard Discoverer spacecraft, but the launching of dedicated sigint types started in 1962.

Two patterns may have been flown: the wide-coverage types designed for "quick-look" sampling of signals, plus more specialized spacecraft optimized for targets of interest. Later designs, octagonal in shape and weighing around 130lb (60kg), could be piggyback launched on the same booster as photo-recce satellites.

Most of these early sigint satellites orbited at heights of 185-310 miles (300-500km). The piggyback models seem to have used a solid-propellant rocket to lift them from the relatively low orbit of the main payload into the higher orbits used for sigint work. After recording radar and other signals, they would transmit data to ground stations for subsequent analysis.

In addition to the dedicated sigint spacecraft, other satellites carried listening equipment. The KH-11 photographic surveillance satellites launched from December 1976 onward flew in 300-mile (480km) orbits and are reported to have carried sigint sub-systems.

Above: The protective shroud drops away from a USAF Discoverer satellite moments before lift-off of its Thor-Agena launch vehicle.

Monitoring microwaves

For years, the Soviet Union assumed that microwave point-to-point radio links operating deep within its territory were immune to monitoring, since these used narrow pencil-beam transmissions accurately directed from one antenna to another. The tiny amount of signal escaping from the main beam in the form of sidelobes would be very difficult for US ferret satellites to intercept, since the low level of signal available in each sidelobe – a thousandth or less of that in the main beam – would be too weak to be gathered by wide-coverage antennas on the relatively fast-moving US ferret satellites.

The gain of an antenna is in part proportional to its size, so it was in theory possible for the US to place into the 22,300-mile (35,880km) geosynchronous orbit a ferret satellite with a massive antenna whose narrow high-gain beam could be accurately aligned to capture faint signals leaking via sidelobes. Since this would have involved orbiting what amounted to a medium-sized radio telescope, Soviet engineers

Microwave interception

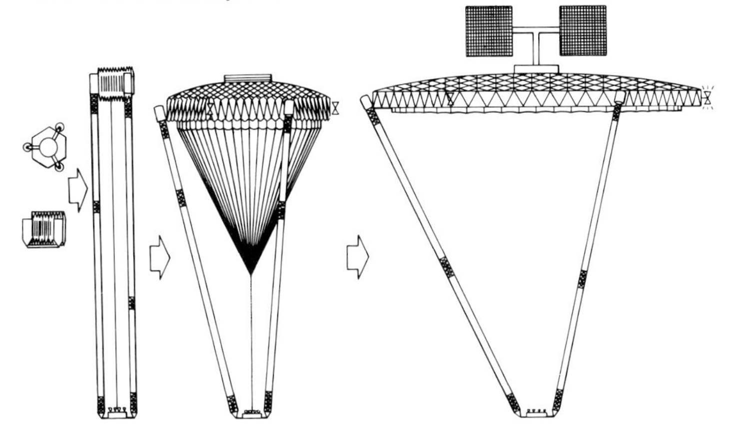

Above: The Rhyolite satellite has a very large antenna dish. One possible technique for creating such a large structure involves orbiting the antenna in packaged form (left). Telescopic booms are then extended, and the backing for the reflector is deployed, while lanyards hold the thin membrane which will form the reflector clear of the expanding truss structure. Once the backing structure is deployed, electrostatic charges are used to draw the membrane toward the backing, stretching its surface into a paraboloidal form. The booms hold the antenna feed at the focus of the newly-created dish.

Right: Rhyolite's massive dish has the gain needed to detect the tiny amounts of energy which leak from the narrow beams of microwave communications links operating deep within the Soviet Union. Such faint signals were immune to interception by the low-gain antennas on conventional low-orbiting ferret spacecraft.

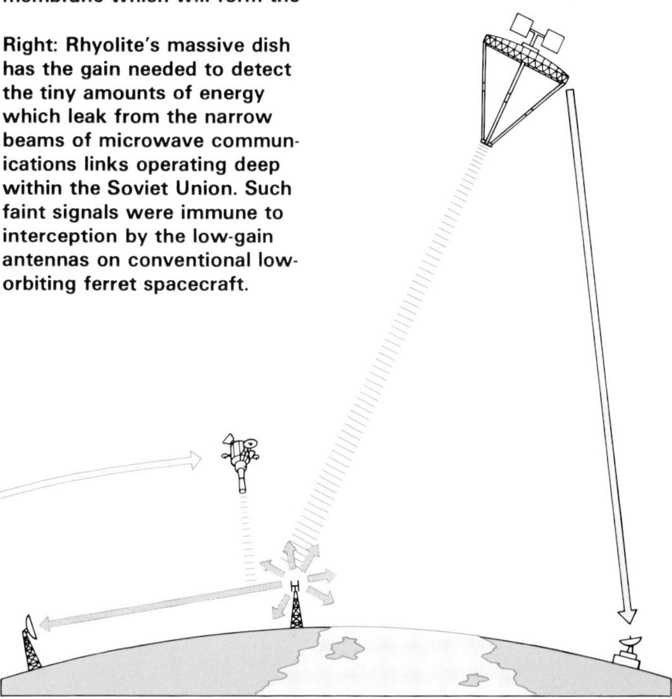

assumed that the technology to create and build such a dish did not exist. The microwave links were therefore secure, and careful timing of missile trials held deep within the Soviet Union, to ensure that no ferret satellite was passing overhead, would ensure that those too were immune to monitoring.

Unfortunately for the Soviet government, they were wrong: the technology to build large antennas in orbit did not exist, but the US National Reconnaissance Office was prepared to foot the bill for creating it. The resulting Rhyolite spacecraft was designed by TRW Defense and Space Systems and assembled in the highly-classified High Bay area of that company's plant at Redondo Beach, California.

Rhyolite in orbit

Rhyolite was a massive spacecraft. Described as "half the size of a freight car" it probably went into orbit aboard a Titan 3. The first was orbited on December 20, 1972, and manoeuvred into a position over the Horn of Africa.

Exactly how the designers tackled the task of creating the collapsible 70ft (21m) dish antenna which Rhyolite used for signal gathering has never been revealed, but a clue may perhaps be found in research ordered by the US Defense Advanced Research Projects Agency (DARPA) and Ballistic Missile Defense Advanced Technology Centre. This showed that large antennas up to diameters of 230ft (70m) or even 656ft (200m) could be created using electrostatically controlled membranes mounted on light circular lattice structures. Once the latter had been deployed to create the circular shape of the antenna, electrostatic charges could be applied to the membrane, drawing it into the desired dish-like shape.

From its monitoring position 22,300 miles (35,880km) up, the spacecraft would be able to observe targets such as military command centres in the Western USSR, plus the Tyuratam and Plesetsk missile ranges. Data was passed to ground stations at Menwith Hill in England and Pine Gap in Australia. Classsified with the special "Tango Kilo" grading, the intercepted signals were then studied by experts of the NSA and GCHQ.

A second Rhyolite was launched in 1973 although the claimed launch date of March 6 given by one source seems questionable. The only payload launched on that date was what seems to have been an early-warning satellite orbited from Cape Canaveral by an Atlas Agena. This second Rhyolite was positioned over Borneo to monitor China and the Eastern USSR (including the highly classified test range at Sary Shagan), while two more were orbited as spares.

Rhyolite seems to have been an interim design. Plans for the larger Argus spacecraft equipped with a 140ft (43m) antenna were drawn up in the mid-1970s, only to fall victim to post-Vietnam cuts in defence expenditure. With the collapse of the Iranian monarchy and the loss of sigint sites in that country, the US government faced a crippling shortage of facilities able to monitor Soviet missiles. The loss of Argus was keen-

Left: A Titan IIID booster lifts off on its way to orbit the first of the USAF's Big Bird reconnaissance satellites.

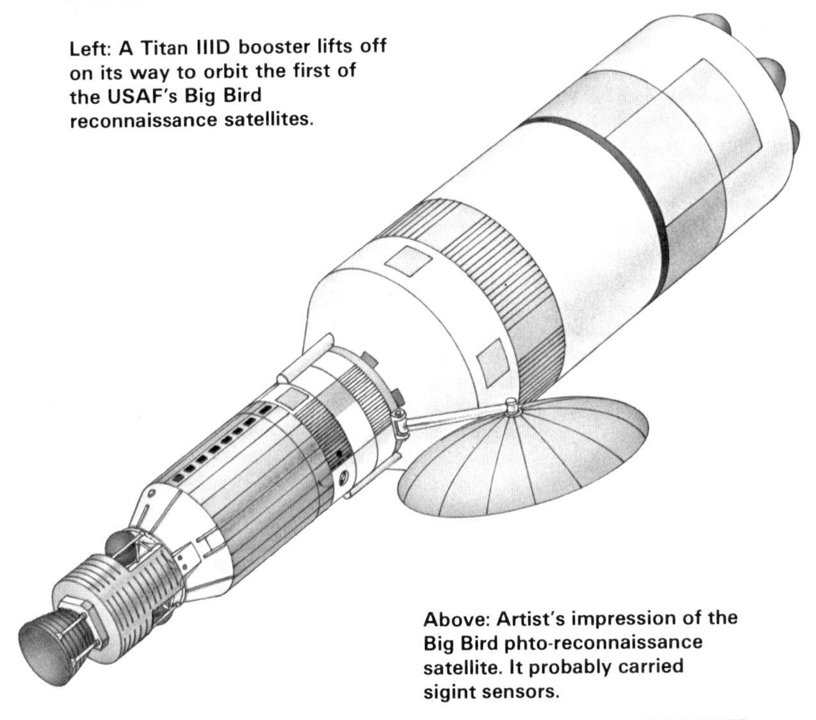

Above: Artist's impression of the Big Bird phto-reconnaissance satellite. It probably carried sigint sensors.

ly felt. A prototype had even been orbited on June 18, 1975 before the project was shelved, and some of its technology may have been hastily embodied in existing designs.

Betrayal

By that time, luck had run out on the Rhyolite and Argus concepts, details of these spacecraft and their unique capability having been passed to the Soviet Union by two traitors working independently in the USA and UK. In the summer of that year NSA employee Christopher Boyce began passing information on the spacecraft to the KGB, while in the UK CGHQ's Geoffrey Prime (who had been spying since the late 1960s) passed to the Soviets highly classified data gathered by these craft.

In January 1977, Boyce's courier was arrested in Mexico while attempting to pass microfilmed data to his Soviet contact, but Prime was to retire from GCHQ in 1977 with his espionage activities still undetected (he was not unmasked until his arrest for sexual offences in 1981). In the meantime, the Soviet Union attempted to counter both the Rhyolite/ Argus type spacecraft and the ground-based listening stations in Turkey and China by encrypting the telemetry transmissions from missiles under test.

Aquacade and Cosmos

Follow-on to the abandoned Argus project was Aquacade. This was a much heavier spacecraft designed to be orbited by the NASA's Space Shuttle. Problems with the Shuttle repeatedly delayed the launch of the first example until December 1984 when it achieved short-lived notoriety after the DoD insisted on a news blackout on the Shuttle's Aquacade cargo while the vehicle flew its first military mission.

Like Soviet photo-reconnaissance satellites, Soviet sigint spacecraft are assigned Cosmos designations. In the early 1970s the Stockholm International Peace Research Institute identified a number of spacecraft launched from Plesetsk as probable electronic-reconnaissance types.

Countermeasures

No matter how rarely an electronic system radiates, the waiting elint ears will eventually detect it. Modern techniques such as frequency agility and spread-spectrum modulation, intended to reduce the "visibility" of the signal to hostile receivers, will make the detection and monitoring of such signals a difficult task. Once intercepted by sigint receivers, signals of this type may be recorded then subjected to detailed frequency analysis in an attempt to discover their characteristics. If necessary, specialized receiving equipment can then be designed to facilitate future monitoring.

Millimetre-wave emitters can create problems for elint system designers. A single receiver can cover the whole of the widely-utilized I and J bands, but coverage of the entire millimetre-wave spectrum requires up to four receivers. The narrow beams associated with millimetre-wave and laser signals makes the design of elint receivers for such transmissions difficult. A "near-miss" by a radar beam will provide sufficient energy to trigger a normal sigint receiver, but this is not the case with these very narrow beams.

Another problem is that the relatively short range of laser and millimetre-wave sensors makes long-range elint operations almost impossible. Much of the earliest data on Soviet work in these fields came from highly classified clandestine sources rather than conventional elint operations.

Below: An operator sets up the Plessey ICE (Interference Cancellation Equipment). Designed to reduce the effects of enemy jamming, it allows users to deliberately jam their own communication channels, blinding enemy sigint systems.

Plessey Interference Cancellation Equipment (ICE) operation

Below: Given the all-round coverage of a normal VHF antenna, a receiver is swamped by the unwanted jamming signal (shown in red), and cannot receive the wanted signal (shown in blue).

Bottom: ICE-equipped receiving stations use two antennas, which allow the coverage to be shaped, reducing sensitivity in the direction of the jammer, and boosting reception of the wanted signal.

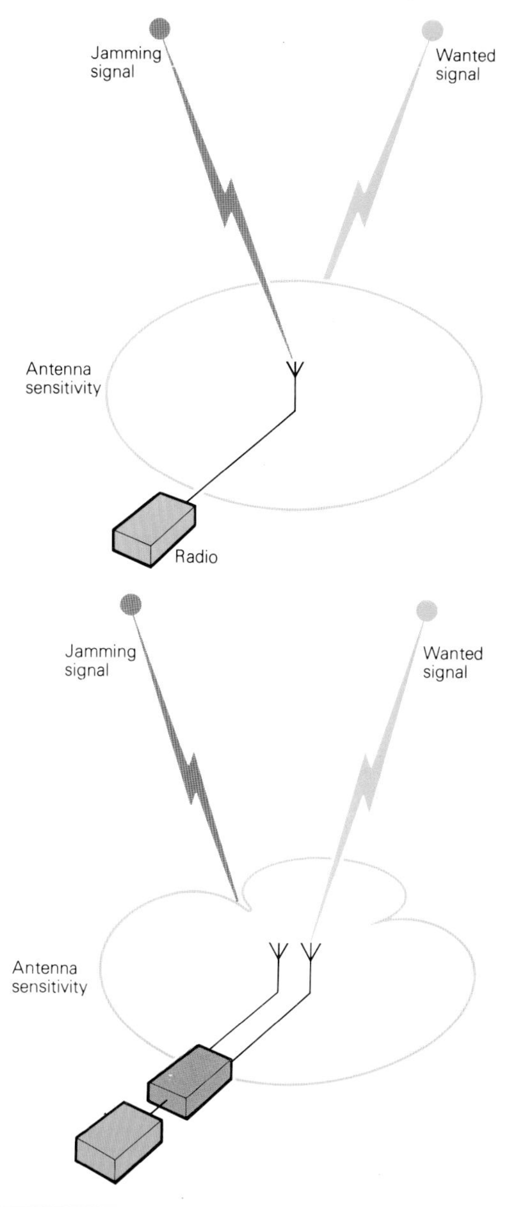

Warning Receivers and ESM Systems

IN THE middle and late 1970s an annual exhibition of military electronics was held in the West German city of Wiesbaden. It did not last, being hounded out of existence by the heavy-handed tactics of German and West European peace groups, but during its brief life it served as a showcase for EW systems.

At one of these exhibitions a senior technical officer of a Middle East air arm, one better noted for its Islamic zeal than for its ability to fly and maintain modern aircraft, visited the display stand of a well-known US manufacturer of EW systems. Company salesmen swung into action, starting a well-rehearsed sales pitch for their principal item of equipment, only to be interrupted by their potential client.

"How powerful is it?"

Below: Soviet-made "Fan Song" target acquisition and fire-control radar at an Egyptian SA-2 "Guideline" missile site.

"What do you mean, Sir?"

"All I want to know is whether it can cope with the Zionist threat – how much jamming power does it have?"

Embarrassed salesmen had to explain that the device in question emitted no jamming power at all. It was in fact an example of the most basic item of equipment – a radar-warning receiver (RWR).

Designed to search for signals from hostile radars, RWRs can detect emissions from surveillance and tracking radars, air-interception radars, and the command links used to guide many patterns of missile. Once the signal has been identified, aircrew are given a warning signal, plus an approximate indication of the bearing, frequency and threat category.

First developed during World War II, RWRs originally were used only in bombers. Early air operations over North Vietnam in the mid-1960s saw US fighters and tactical aircraft ex-

Above: Early attacks on US warplanes by North Vietnamese SA-2 sites such as this newly-constructed battery spurred the development of the first US tactical RWRs.

posed to Soviet-style defences, and the resulting losses shook the USAF and USN out of their traditional complacency and misguided confidence in aircraft survivability, triggering off a crash programme to develop suitable countermeasures.

Countering "Guideline"

When the North Vietnamese defences first used SA-2 "Guideline" surface-to-air missiles against USAF and US Navy fighters, the effectiveness of these first-generation radar-guided weapons must have delighted Soviet advisers watching the engagements. According to USAF figures, half the rounds fired downed their targets, the first victim being an F-4 shot down in July 1965. Within a month the US DoD started work on two methods of dealing with the threat, a measure hastened by the loss over Dienbienphu of a US Navy RA-5C Vigilante to SA-2 fire.

One approach saw the decision to develop specialized anti-radar strike aircraft, a type of weapon described later in this volume; the other, ordered into crash production under the designation APR-25, was a compact RWR called Vector IV. Developed by a small company called Applied Technology — a specialist in the construction of strategic reconnaissance and elint equipment — this used four wide-angled broadband antennas to determine the approximate direction of radar signals reaching the aircraft, and displayed the resulting bearing on a cockpit-mounted CRT display.

Detailed information on the threat was minimal, so the APR-25 simply monitored a range of frequencies and displayed the signals it detected. The technique was simple, but it worked. A companion equipment designated APR-26 was designed to monitor the frequencies used by the SA-2 and its associated "Fan Song" radar when a round was in flight. This would give aircrew warning of missile launches.

Threat detection

The basic principle of RWR operations and tactical use relies on the fact that when a threat radar is first detected by the RWR, the threat may not yet have detected the incoming aircraft. To the layman, it might seem that if the RWR can detect the

Soviet interceptor and missile threats to Western aircraft

Radar band / Frequency	A below 250MHz	B 250 to 500MHz	C 500MHz to 1GHz	D 1-2GHz	E 2-3GHz	F 3-4GHz	G 4-6GHz	H 6-8GHz	I 8-10GHz	J 10-20GHz
MiG-21 FISHBED	Radio & data link								SPIN SCAN AI radar	JAY BIRD AI radar
MiG-23 FLOGGER	Radio & data line									HIGH LARK AI radar
MiG-25 FOXBAT	Radio & data link								FOX FIRE AI radar	
Su-15 FLAGON	Radio & data link								SKIP SPIN AI radar	
Tu-128 FIDDLER	Radio & data link								BIG NOSE AI radar	
SA-1 GUILD					YO-YO Acquisition radar GAGE Tracking radar					
SA-2 GUIDELINE	SPOON REST target acquisition	command link			FAN SONG B missile/target tracking		FAN SONG E missile/ target tracking			
SA-3 GOA			FLAT FACE or SQUAT EYE target-acquisition	Command link					LOW BLOW missile/tgt tracking	
SA-4 GANEF					LONG TRACK target-acquisition		PAT HAND missile/ target tracking			
SA-6 GAINFUL			FLAT FACE target acquisition		LONG TRACK target acqusition		STRAIGHT FLUSH target acqusition		target tracking command link	
SA-8 GECKO								LAND ROLL target acqusition		LAND ROLL missile/tgt tracking
SA-9 GASKIN										GUN DISH target tracking
SA-N-1 GOA				Command link	PEEL GROUP high-altitude missile/tgt tracking			PEEL GROUP high-altitude missile/target tracking		
SA-N-3 GOBLET							HEAD LIGHT target-tracking	target & missile tracking		
SA-N-4								POP GROUP target acquisition		POP GROUP missile/tgt tracking

Note: Elint data on newer systems such as the MiG-29 "Fulcrum", Su-27 "Flanker" and SA-10, -11, -12 and -13 SAMs is still classified.

threat, the threat should be able to detect the RWR-equipped aircraft, but this is not the case. In practice, the RWR has a significant performance edge, since it is attempting to detect signals from a powerful radar transmitter whose peak power may be measured in kilowatts — or even megawatts in the case of many types of long-range surveillance set. Although the RWR has only a low-gain omnidirectional antenna array, the power of the incoming signal is sufficient to make detection relatively simple.

The radar has a more difficult task, since the amount of energy available to its receiver is smaller by several orders of magnitude. The signal utilized by the RWR must be reflected from the target — losing energy in the process — then travel all the way back to the radar antenna. Despite the latter having a large and highly directional antenna, it has little chance of detecting the faint echo signal before its own powerful beam has triggered the target's RWR. Given this warning, the pilot may change course to avoid the radar, reduce cruise height in order to fly beneath the radar's horizon, or turn on some form of ECM. The warning provided by the RWR gives him a clear advantage.

If a surface or airborne radar manages to lock on to the aircraft, this will register on the RWR display. Pilots attempting to break the lock of threat signals by deploying chaff or flying very low will be able to determine whether their tactic was successful by watching the RWR output to see if the hostile signal is still present.

RWRs fitted to Western military aircraft do not need to monitor the entire range of frequencies used by contemporary Warsaw Pact anti-aircraft systems but include at least one operating frequency of each surface-to-air missile system. To take an example, the SA-3 Goa missile has a C-band acquisition radar and D-band command link but will betray itself to the British ARI series receivers by the I-band signals used for target and missile tracking.

Above: The Soviet SA-N-4 missile (seen here being launched from a Koni class corvette) is guided by the H-band "Pop Group" radar.

Early Western RWRs

The APR-25 and -26 were rushed into production and into service and had an immediate effect on aircraft loss rates. Within months, a combination of RWRs and Wild Weasel anti-radar strikes brought the kill rate down to only three per cent, with 30 "Guidelines" being launched for every USAF fighter downed. The APR-25 and -26 were soon supplemented by the later APR-36 and -37, improved RWRs with additional signal-processing features.

Similar equipments were soon developed by other nations, such as the Marconi Space and Defence ARI 18223 fitted to Royal Air Force single-seat fighters such as Jaguar and Harrier. Forward- and rearward-facing antennas for the system are carried in rectangular housings which are conspicuous features on the vertical fin of these aircraft. Operating frequency is probably E-J band, and information is presented to the pilot by lamps on a display unit which indicates the frequency band and nature of any signal received, and the sector from which it is coming.

This simple lamp display can only give details of a single threat at any one time, but the CRT display of the ARI 18228 fitted to multi-seat RAF aircraft such as the Phantom and Buccaneer can show several at once, indicating bearing, transmission type and other data. This equipment uses the same RF sub-systems as the ARI 18223.

Soviet systems

The Soviet Union developed its own RWR systems, known collectively as

Sirena, and these have evolved through several models. The early Sirena 1 and 2 gave only rear-sector coverage, but the SO-69 Sirena 3 fitted to most Soviet fighters gives all-round coverage. Forward-facing antennas for this equipment may be seen on the leading edge of the fixed glove section on the MiG-23 "Flogger"; the rear-sector antenna is located in a bullet fairing near the top of the vertical fin trailing edge. On the MiG-25 "Foxbat", the Sirena 3 antennas are located in the wingtip anti-flutter pods.

Early Su-7 "Fitters" had no RWR, but the Sirena 3 was introduced on the Su-7BMK version during the mid-1960s. This equipment is also a standard fitting on the Soviet Air Force Su-17 variable-geometry "Fitter" variant. Export Su-22s have simpler avionics, which includes the Sirena 2, but there are no details of the RWR supplied with Su-20 "Fitters" serving with non-Soviet Warsaw Pact air arms. Sirena 3 is also a standard fixture on all models of the Su-15 "Flagon".

The RWR carried by the Su-24 "Fencer" has not been identified, but is probably a more recent design than that fitted to earlier aircraft. The limited signal-processing capability of the mid-1960s vintage SO-69 Sirena 3 receiver is hardly likely to meet current demands, so a more modern equipment may be assumed. Such advanced RWRs may even have been retrofitted into other Sirena-equipped aircraft.

Sirena 3 is also fitted to Soviet bombers. On the Tu-95/142 "Bear", the antennas are probably located in the pods at the tips of the horizontal tail surfaces. Analysts have recognized Sirena 3-style antennas on the Tupolev Tu-22M "Backfire", but once again it is dangerous to assume that these feed the mid-1960s vintage SO-69.

New techniques

By the late 1960s, new anti-aircraft threats were being fielded in Vietnam. The growing number of emitters, and the increasing use of new modulation methods by Soviet equipment in North Vietnamese service, forced a change from the

Right: The forward-facing antennas of the SO-69 Sirena 3 radar-warning receiver are mounted in the leading edge of the MiG-23's wing glove; the aft-facing antenna is located in a bullet-shaped fairing on the trailing edge of the vertical fin.

Right: On the Saab-Scania Viggen, a forward-looking RWR antenna is mounted in a bullet-shaped fairing which forms part of the vortex-generating dogtooth on the leading edge.

RWR installations

Below and left: RWR antennas are designed to cover a wide angle, and a broad range of frequencies. One successful design is the spiral antenna shown below — the tip of the spiral handles the highest frequencies, while the aft sections deal with progressively lower frequencies.

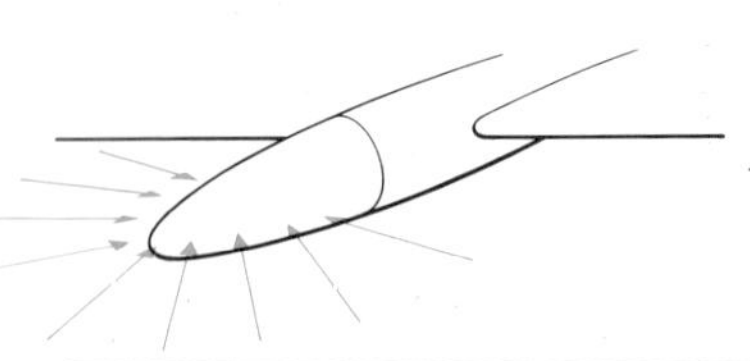

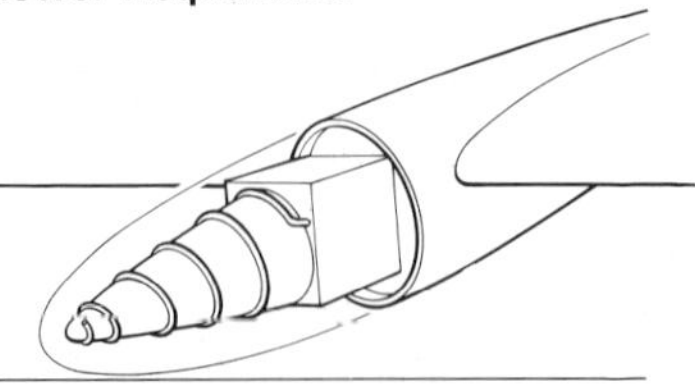

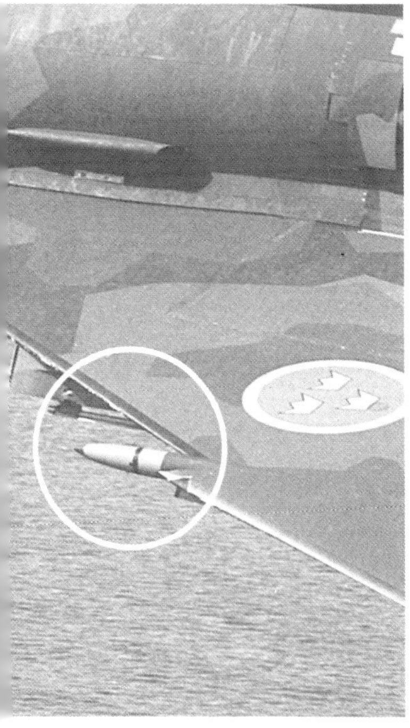

"signal-grabbing" tactics of earlier receivers.

If an RWR warned the aircrew of every single signal, they would soon switch it off to terminate the endless false alarms. RWRs intended for use in the high signal density had to incorporate a degree of signal processing so that only those signals representing a threat to the aircraft would be displayed. In the development of new RWRs such as the US Navy's ALR-45 and the USAF's ALR-46 much importance was therefore placed on identifying the individual emitters, discarding non-lethal threat data, and displaying the highest priority threats.

Threat classification

In the earliest stages of an attack, long-range surveillance radars will be a threat, but only a long-term one, since the longer the time the enemy has to track the intruder, the more time he has to prepare his defences. Aircrew may wish to know that they are being illuminated by such radars, but might not wish an alarm signal unless the radar starts tracking them — a reliable sign that they may soon be engaged by the defences. If signals from lock-on missile tracking or air-interception radars are detected, this represents an immediate threat, since attack by missiles or interceptors may be imminent. Signals from surveillance radars are of no further interest at such a time — the direct threat comes from defensive systems attempting to destroy the aircraft.

In order to minimize aircrew workload in combat, all these analysis functions had to be carried out automatically, which meant using digital control circuitry in place of the analogue electronics in earlier receivers. Threats were identified by comparing parameters such as frequency, pulse width, PRF and scan pattern with a built-in catalogue of threat parameters compiled from the results of electronic intelligence-gathering (elint) operations. Other changes in signal parameters (such as a change in modulation or power level, or the presence of missile-guidance data link signals) can warn the pilot that a missile has been launched. The ALR-46 was the first software-programmable RWR to enter service.

Tracking indication

With older types of radar it is easy for a target to detect that it is being tracked. So long as the threat radar continues its regular scan pattern and its signals are received only at intervals as the target is briefly illuminated by the moving radar beam, then all is well — the threat radar may be monitoring the position of the target, but has not singled it out for special attention. If the signal suddenly becomes continuous, this indicates that the threat radar has ceased scanning and is now concentrating its attentions on the target.

This simple and comfortable rule does not apply to track-while-scan (TWS) radar threats, since the latter can track the target while continuing their regular scan pattern. All TWS signals must therefore be identified as potential threats, a task which can be carried out only by comparing the parameters of all signals detected with a threat library.

One non-US RWR with this capability is the Thomson-CSF Type BF, whose forward- and rearward-facing bullet radomes may be seen on the tail surfaces of the Mirage F1 and Super Etendard; two flush-mounted spiral antennas provide coverage of the left- and right-hand sectors. Audible warning of a radar signal is given to the pilot, while cockpit-mounted display lamps show the general direction and nature of the threat. No frequency indication is given, but the system will distinguish between pulse radar and CW or ICW radar, and will warn the pilot if the set is of the TWS type.

Threat proliferation

More recent developments in RWR technology have expanded the frequency coverage to cope with newer

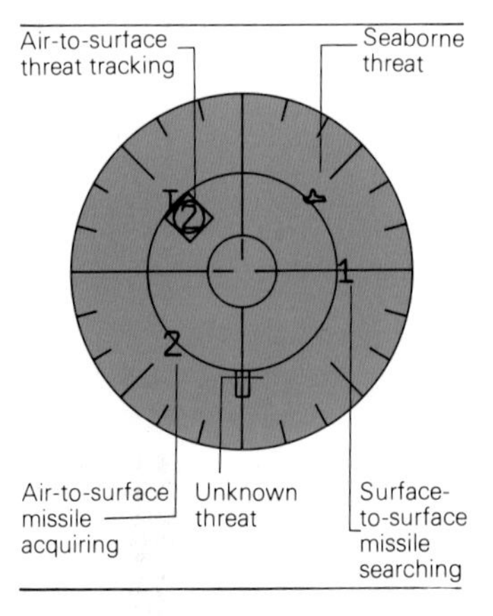

Above: Display of the Dalmo Victor Triton naval RWR showing typical symbology used to designate threat signals.

Below: Blade and spiral antennas, avionics units and cockpit display and control unit of the MEL/Dalmo Victor Katie RWR, designed for use by helicopters and light fixed-wing aircraft.

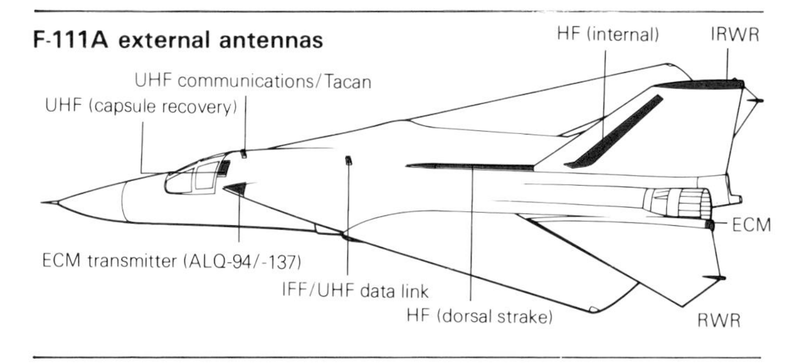

Top: The F-111 carries sensors and antennas in abundance; its Warsaw Pact equivalent — the Su-24 "Fencer" — seems near-naked by comparison.

Above: RWR and ECM antennas are located on either side of the F-111 jetpipes and at the fin top.

threats and increased the signal-processing capability. Even in a limited conflict such as the 1973 Middle East War, the Egyptian Army was able to deploy more than 100 batteries of SAM systems along the Suez canal. Allowing for the fact that each battery could be simultaneously transmitting several different radar or command-link signals, an aircraft flying along the length of the canal could expect to detect up to 500 signals.

On the NATO central front, the problem would be even more severe. The older SAM systems and radars used by Egypt have been supplemented but not totally replaced by newer equipments, while radars associated with NATO SAMs and interceptors of both sides would add to the electronic cacophony. It is realistic to assume that thousands of signals would be detected.

Needless to say, the individual pulses of radar energy from all these radars do not arrive in neat and regular order like well-drilled soldiers on parade but in an overlapping chaos which makes rush hour at a railway terminus seem simple by comparison. In any realistic situation, more than 100,000 pulses will be arriving at the receiver antenna every second, friend and foe, threat and non-threat mixed into an electronic mélange which must be sorted out and categorized. Conventional electronic technology simply will not be able to cope with future threat levels, so electro-optical devices such as Bragg cells are being investigated as potential solutions for future RWR and ESM equipment.

ESM Systems

As RWRs become more complex, they can no longer be considered simple warning devices, but must be categorized as electronic support measures (ESM) – equipments with much greater signal processing power able accurately to locate and identify threats and even to initiate action against them.

Western Europe has produced airborne ESM systems, but few details are available of the equipment fitted to the Panavia Tornado. Developed by AEG-Telefunken, Elettronica, and Marconi Space and Defence systems, the system carried by strike Tornados uses forward- and aft-facing antennas mounted in a fairing near the top of the vertical fin. Royal Air Force Tornado F.2 (ADV) interceptors carry a Marconi Defence Systems ESM system incorporating extensive digital signal processing and LED cockpit displays. Using this equipment, the aircrew will be able to detect and home in on radar-emitting targets.

Some receivers directly control the aircraft's ECM systems, deploying chaff or flares, switching jammers on and off and adjusting the frequency of the output signal. One example is the Loral ALR-56 carried by the F-15 Eagle. In addition to providing audible and visual warning to aircrew, this complex receiver system directly controls the aircraft's ALQ-135 ECM equipment.

In order to cope with the wide range of frequencies which the Eagle might have to face, the system is divided into two sections. Four spiral antennas mounted at the tips of the wings and vertical tail surfaces handle the higher frequency signals, passing these to the "high-band" tuner – a dual-channel receiver able to carry out direction-finding operations in order to determine threat bearing. Lower-frequency signals are received by an omnidirectional blade antenna located on the lower fuselage. Signals from this unit are passed to a simpler single-channel "low-band" receiver contained in the same avionics package as the computer used for receiver control and signal processing.

All detected signals are passed to a pre-processor, which analyzes their characteristics, then passes the threat data in digital form to a computer which selects data for display and transmission to the Eagle's jammers.

Signal analysis

Surface ships and large aircraft such as bombers and maritime-patrol types are good platforms for ESM systems, having the space, electrical power and personnel to handle systems able to intercept, categorize, sort and identify all detected signals. Once threats and known "friendlies" have been identified with the help of a built-in library of emitter signatures, and the latter have been electronically "tagged" to be ignored, threats and "unknowns" can be displayed, the operator receiving only the high-priority data which he really requires.

Like RWRs, ESM systems work in real time. Analysts of elint data can

take as long as they like to study an intercepted signal, but in times of crisis or combat the ESM operator needs data immediately. Frequency, PRF and pulse rate may be quickly measured, and are sufficient to identify most emitters, but in some cases the scan pattern must also be analyzed — a process which take time, but which might provide the final clue. An unknown radar which suddenly switches from general surveillance to tracking the ESM-equipped ship or aircraft is unlikely to be friendly.

Surface systems

ESM systems can be ship- or land-based. Free of the very strict space and weight constraints which apply to airborne systems, ship- and particularly land-based ESM systems may be as complex as technology and the customer's pocket will permit. A typical shipboard system is the Racal Radar Defence Systems RDL Series of manually-operated ESM systems. Adopted by 18 navies, this

Above: Radar-frequency antennas of the F-15's ALR-56 ESM set are on the fins (1) and wingtip (2); belly-mounted blade antennas (3) cover the lower bands.

is available in forms ranging from the basic E-I band RDL-1BC for patrol craft up to the RDL-5, -6, -7 and -8 which cover from D-J band (1-18 GHz). If linked to an SRU-1 signal recognition unit, an RDL system can compare the parameters of the unknown signal with a library of up to 500 known emitters.

Being manually operated, the RDL series can be overwhelmed by the sheer number of signals in high-threat areas, but the same company's Cutlass computer-controlled system can tackle signal densities of up to 500,000 pulses per second over the frequency range of D-J-bands. Signals are automatically compared with the built-in threat library and the operator is then presented with a tabulated listing of their identities

and degree of threat. The 25 highest-priority threats are displayed automatically, but the operator can request the next 125 in descending order of threat priority. Cutlass does not use sweeping techniques to monitor the range of frequencies but uses an instantaneous frequency measurement (IFM) receiver, thus giving a high degree of probability that a signal of short duration, such as that produced by a strike aircraft turning on its radar for a "quick-look" tactical update, will be detected.

Land-based systems intended for army use are often mobile. A typical system such as the US Army's TSQ-109 AGTELIS (Automatic Ground Transportable Emitter Location and Identification System) is mounted in 2.5t trucks so that it can be transported to forward positions and used to search for emitters operating between 500kHz and 18GHz. Data from three GSQ-189 outstation vehicles measures the frequency, pulse width, PRF, and other parameters of hostile signals, passing the resulting data to a two-vehicle signal processing facility. The TYQ-17 processor handles the data flow and controls the outstations, while the TSQ-11 processor integrates this data and passes it to a control and analysis centre.

Millimetre-wave systems

Most current ESM systems cover up to 18 or 20 GHz, so can easily cope with J-band systems such as the "Gun Dish" radar of the Soviet ZSU-23 Shilka self-propelled anti-aircraft gun, or the "Jay Bird" radar of second- and third-generation MiG-21 "Fishbed" fighters. Soviet developments into the millimetre-wave region of the spectrum have already spurred the development of millimetre-wave ECM techniques by the US and UK, while Western equipments such as the Marconi

Above left: The add-on millimetre-wave circular antenna of this Contraves Sky Guard fire-control radar improves low-level tracking performance.

Above: When Signaal's millimetre-wave Flycatcher fire-control radar was tested in the UK, British engineers were amazed by its low-level capability.

Blindfire and Signaal Flycatcher radar trackers are probably prompting similar efforts by Russian RWR designers.

IR warning

Infra-red warning receivers can be used to detect the launching of tactical missiles, enabling ECM systems to be turned on or evasive tactics to be carried out. This is particularly valuable if IR jammers are to be used, since the life of the lamp sources used in many systems is limited. Early IR receivers suffered a high false-alarm rate, but newer designs are reported to be more reliable.

Counter-ESM techniques

One counter to RWR and ESM systems is the use of frequency agility. Since the emitter no longer occupies a single discrete frequency it is harder to monitor and categorize. In order to cope with radars which use this technique (including some recent Soviet designs), elint efforts are being focused on identifying the electronic equivalent of fingerprints, tiny but unique features of the signal which will allow the equipment to be identified. These might include the slope of the rising and falling edge of the pulse shape, or ringing or ripple — minor but regular flaws in the pulse shape. Consistent for any radar, these should allow even the latest frequency-agile sets to be detected and recognized.

Another obvious counter is to

move to operating frequencies above the coverage of existing sets, using millimetre-wave or even laser-based guidance. Reports of tiny radomes that could only house millimetre-wave antennas on new Soviet weapon systems and platforms (including tanks), and the presence of fairings for electro-optical sensors or lasers on recent versions of fighters such as the MiG-23, confirm that this is in fact a tactic which the Soviet Union has chosen to adopt.

Expanded coverage

The counter to these must be new patterns of Western RWR and ESM equipment intended to cover the new frequencies. The US Advanced Laser Warning System will provide a complete coverage of the optical spectrum. A warning system for the Detection of Laser Emitters (DOLE) has also been developed and flight tested.

In 1979 Dalmo Victor and Perkin-Elmer demonstrated to the US and NATO military a multisensor warning receiver designed to cover conventional radar, millimetre-wave and laser frequencies. On test, this detected threats in the 20-40GHz region of the spectrum, responding to laser beams but correctly rejecting high-intensity beams of light from non-laser sources.

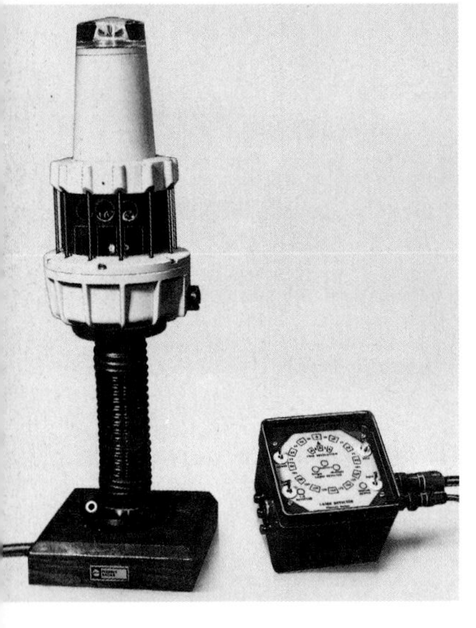

Above: IR warning receivers have often proved unreliable. The USAF decided to delete the rear-facing ALR-23 sensor (the dark object on the fintop fairing) planned for the Grumman EF-111A before the first example was deployed to the UK.

Above left: This Plessey warning receiver for armoured fighting vehicles can detect emissions from battlefield lasers and IR searchlights, warning the crew that they are under threat.

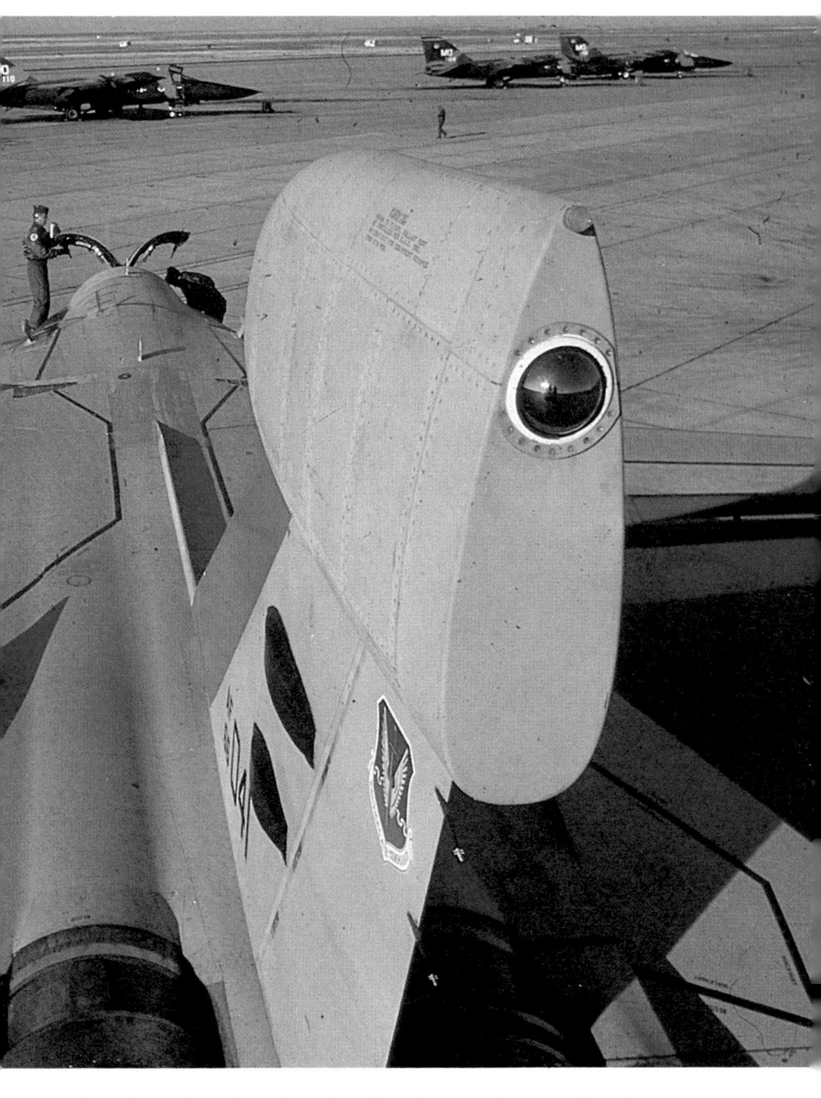

Experimental work continued, and in 1982-83 the US DOLRAM (Detection of Laser, Radar and Millimeter Waves) programme saw the flight testing of an integrated warning suite incorporating a laser warning receiver, angle-of-arrival receiver and a millimetre wave receiver. Currently, under project F34-375, an Integrated Tactical Electronic Warfare System (ITEWS) is being developed for use aboard USN aircraft. This advanced system will include electro-optical, infrared and ultraviolet technology for target detection and classification, and may also incorporate facilities for targeting weapons against detected threats.

Taking the frequency-agility principle further, spread-spectrum modulation will make future transmissions more difficult to detect. Although its adoption will increase the cost of the emitter, its effect on the cost of the ESM system will be even greater. In the long run, the use of spread-spectrum techniques might even severely erode the technological edge which currently favours the RWR or ESM system.

EW 'Expendables'

Chaff, Flares, Smoke & Decoys

COMPARED with active jamming, the use of expendable EW weapons such as chaff, flares, smoke and other decoys might seem crude, but there is no denying the cost-effectiveness of such techniques. Proven in combat in the Middle East and the South Atlantic, expendables are now an EW aid which few armed forces dare ignore.

Most of the threats in modern warfare are vulnerable to attack in this manner. Chaff will deal with radars and radar-guided weapons, flares can be used against IR-guided missiles and tracking systems, and decoys can confuse long-range sensors and help ballistic missiles break through ABM defences by multiplying the number of targets.

Chaff

The simplest anti-radar countermeasure is chaff — small strips of conducting material whose length is selected to make them good reflectors of radar energy, the optimum length being half the wavelength of the radar signal being countered. Radar signals induce tiny electrical currents within the individual elements, causing them to re-radiate a signal at the same frequency. Each strand acts as an efficient receiver and re-transmitter of the wavelength to which its physical length is matched, so that its effective size as seen by the radar is much larger than its physical size. The technique dates back to the 1940s but is still useful.

The chaff used during World War II was made from aluminium foil, but modern chaff is made of aluminized glass fibre or silvered nylon fibre. The fact that only the outer layer is electrically conducting does not impose any penalty, since at the high frequencies used by radar, electrical currents flow not throughout the cross-section of a wire, but only near the skin.

Once dispersed into a compact cloud, a small chaff package about twice the size of a pack of cigarettes will match the echoing area of a jet fighter, and being extremely light it will take a long time to drift to the ground. At sea level chaff will fall at only one or two feet per second, and even in the rarefied air at altitudes of 50,000ft (15,000m) will be less than 10ft/sec (3m/sec). As a result, chaff clouds are long-lasting.

Chaff deployment

Several deployment methods are used. When used to protect aircraft, chaff may be released continuously to build up protected corridors through which aircraft may be routed, or behind which units may deploy into attack formation free from radar observation. So low in density is modern chaff that the crew of aircraft flying a chaff corridor will not be able to see it: to make sure they do not accidentally drift out of the protected zone, they will have to use their radar.

One problem with using chaff in this role is its lack of motion. Being light in weight, it has insignificant momentum, so rapidly loses speed after deployment. If the threat radar is fitted with moving-target indication (MTI) circuitry designed to eliminate ground clutter, this feature will be equally effective against near-stationary chaff clouds.

Released in short bursts, chaff can be used to create false targets, or as a self-protection device to break the lock of tracking radars. Most types of target-seeking system direct the

Top right: This radar image, taken from the display screen of a Ferranti Crest radar trainer, shows the effects of a chaff corridor (left of centre).

Right: Individual chaff filaments are so fine — typically around 0.1mm in diameter — that hundreds may share the eye of a needle with a cotton thread.

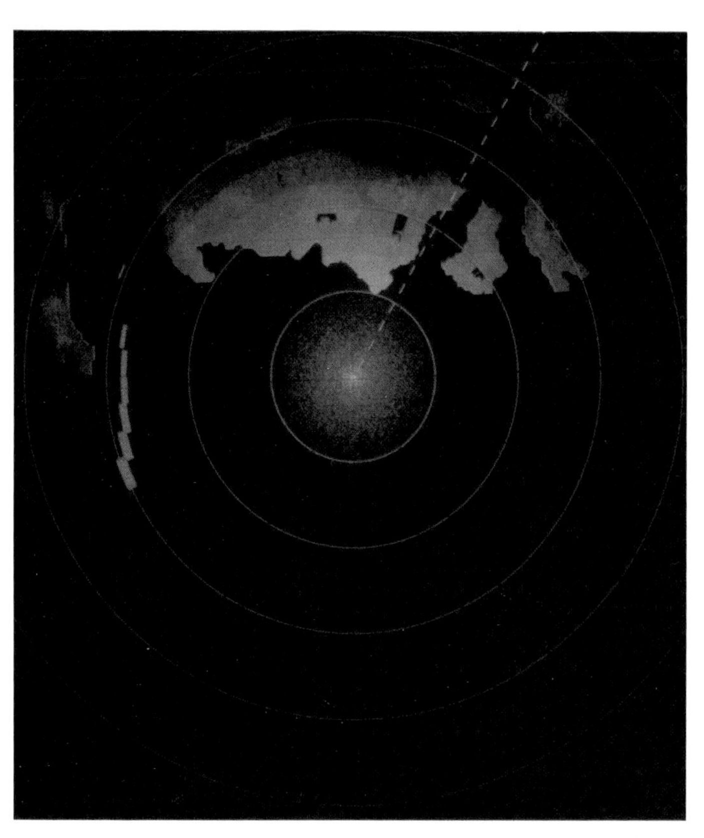

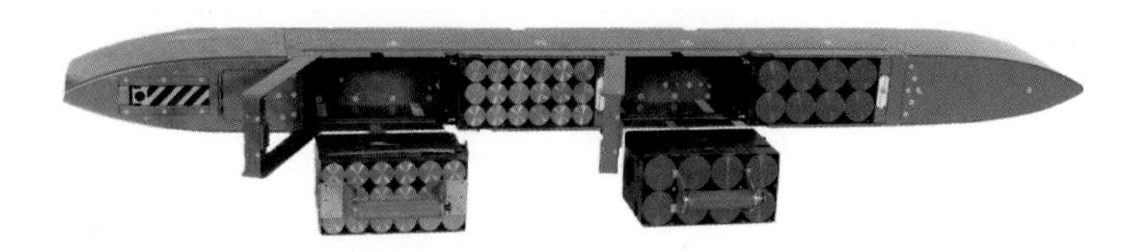

Above: One of the few British products in the field is the Wallop Industries Masquerade system, which neatly packages chaff and IR flare cartridges into a standard CBLS (container bomb light store) type 200.

Top: Alkan dispenser with 40mm chaff cartridges (left) and 74mm infra-red decoy flares (right).

antenna of a radar at the centroid of the radar returns, and chaff clouds are intended to move that centroid away from the parent aircraft. In this case, the rapid drop-off in speed of the newly-released chaff is beneficial, since it will quickly separate the cloud and the aircraft. The tracking radar will almost certainly follow the chaff, particularly if the aircraft executes a sharp manoeuvre immediately after chaff release.

Bloom time — the interval required for a chaff cloud to form — is critical: the faster the platform, the shorter the bloom time must be if the chaff cloud is to be fully formed and drawing the attention of the hostile radar before the platform has left the "field of view" (or, to use the correct engineering term, resolution cell) of the radar beam. If the cloud does not form quickly enough, the radar will still be concentrating on the target as the latter starts to leave the resolution cell and will correct the antenna position to keep the target in view, ignoring the distraction offered by the growing chaff cloud. Typical bloom times for air-launched chaff are around 50 milliseconds.

Chaff dispensers

The simple nature of chaff lends itself to improvized dispensing methods. The trick of stuffing chaff behind the air brakes or inside the wheel wells of an aircraft then momentarily opening these in combat is a useful wartime improvization for air arms which do not have proper dispensers, but is hardly the ideal technique. A slightly more sophisticated trick devised during the 1970s by RAF ground crews at Bruggen in West Germany involved a crude chaff dispenser which could be fitted within the housing on the Harrier normally used to carry the braking parachute.

The normal method of releasing chaff, however, is by means of customized dispensers. These usually take the form of equipments designed to fire cartridges of pre-cut chaff material which burst in the slipstream to form reflective clouds.

One highly effective alternative type of dispenser comprises a pod fit-

Rapid bloom chaff deployment

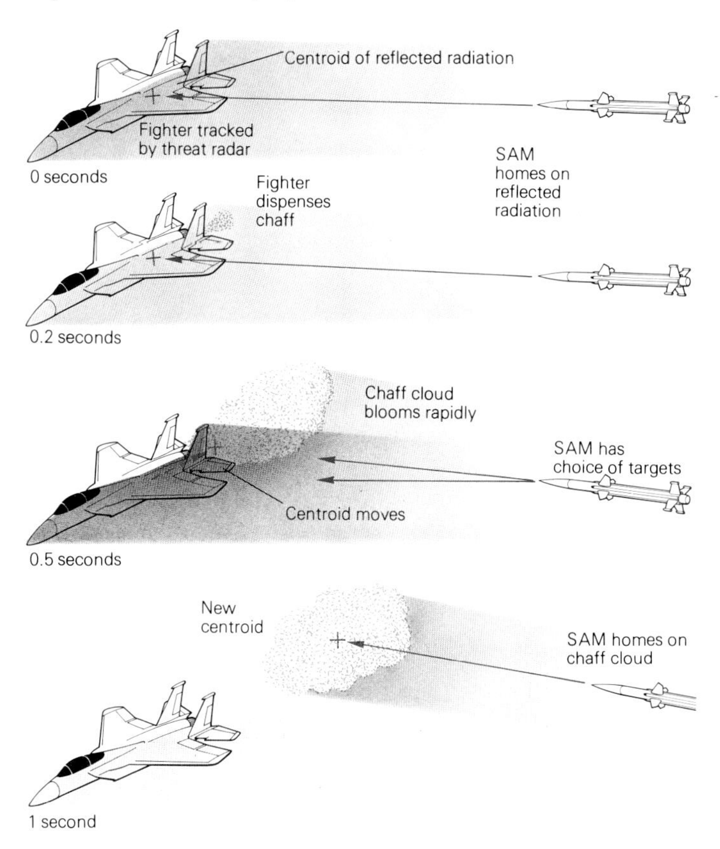

Above: Rapid bloom chaff spreads quickly after release, increasing the radar size of the target. If the chaff cloud and aircraft are still both within the seeker's field of view, the seeker will home on the centroid of their combined target areas. As the chaff cloud slows down, the seeker will follow it, while the aircraft passes swiftly outside the seeker field of view.

ted with drums of continuous chaff material and cutting mechanisms, enabling the chaff to be customized in combat to match the exact frequency of the threat. The 185lb (84kg) Lundy ALE-43 pod carries up to 350lb (159kg) of chaff, in pre-packaged continuous lengths, sufficient for almost three full minutes of operation. These are fed to electrically driven cutting blades which can be set to operate continuously on demand or automatically sequenced off and on to provide bursts of chaff. The chaff may be cut to cover the entire spectrum from A band to K band.

Aircraft installations

Chaff was first used to protect bombers, and still serves in this role. The USAF's B-52G and H strategic bombers are fitted with Lundy ALE-24 chaff-dispensing systems, and the use of underwing rocket pods designed to sow chaff clouds ahead of the aircraft has been reported. Remotely-controlled dispensers for chaff and flares are standard fittings on the Tupolev Tu-16 "Badger", while the "Badger-H" dedicated EW aircraft has ventral chutes for dispensing chaff, flares and even expendable jammers.

Above: A US Navy CH-53D dispenses a salvo of IR flares. These are a more attractive target for heat-seeking missiles than the aircraft's T64 engines.

Similar dispensers are fitted in the landing gear pods of the Tu-22 "Blinder". Other Soviet bombers almost certainly carry internally-mounted chaff dispensers. Pods for chaff rockets are carried beneath the outer wings of the Yak-28 "Brewer-E" EW aircraft.

Dispensers are now common features on tactical aircraft, and systems such as the widely used Tracor ALE-40 family may be internally or externally mounted. On the F-4 Phantom, four dispensing units can be installed on either side of the inboard weapon pylons, while other variants employ two or four skin-mounted, internal or semi-internal dispensers located wherever practicable on the carrier aircraft; typical locations include the rear fuselage or belly of the aircraft. It can operate under manual or automatic control, and payloads may be ejected singly or in pre-programmed combinations, depending on the assessment of the tactical situation.

Smart dispensers

One recent development is the smart dispenser. Fitted with a built-in radar-warning receiver (RWR), equipment of this type provides a self-contained self-protection capability for manned aircraft. Racal Radar Defence Systems has developed an antenna array for its Prophet RWR which would allow this unit to be incorporated into the structure of the Wallop Industries Cascade dispensing pod. The latter is intended for helicopter use, and is normally mounted in pairs, one on either side of the aircraft fuselage. Each carries 24 2¼in (57mm) diameter cartridges comprising a mixture of chaff and flare decoys carried in a ratio of 2:1 or 3:1 as required by the user. The larger Masquerade pod-mounted system is available in a version with built-in Decca RWRs, and carries 48 cartridges. Cartridges are fired in salvoes with anything from 0.1 to 6.0 seconds between shots.

US developments are less advanced. The USAF plans to deploy a smart ALE-47 dispenser, but has yet even to choose a supplier. The equipment which finally enters service will probably be more complex than the British systems, and will be designed to automatically counter air- and ground-based threats.

Infra-red flares

Most dispensers used for chaff can also be used to drop infra-red flares capable of confusing heat-seeking missiles. (Some dispensers can also be used to release miniature expendable jammers but this technique is much less common than the use of chaff or flares.) In addition to protecting tactical aircraft, flares also play a role in protecting strategic bombers, and USAF B-52 and F-111 units are now equipped with a new improved type.

Early IR weapons were very vulnerable to decoy flares, but more recent designs use filters or dual operating frequencies in order to estimate roughly where the peak level of IR output lies. In order to pack a large amount of IR output into a small package, the flare must burn at an extremely high temperature while the genuine target is likely to be a large mass of metal (such as the aft fuselage and jetpipe of an aircraft target) at a somewhat lower temperature.

Since the flare, being a higher-temperature source, will have its peak response at a higher part of the IR spectrum, and a significantly reduced output at lower frequencies, the seeker head can, by means of filters and signal processing, distinguish the real target from the false. If the level of higher-frequency IR energy is greater than or equal to that at the lower frequency, the IR source being observed is almost certainly a flare and should be ignored. A genuine aircraft target will show greater output at the lower frequency.

Despite these limitations, flares are still widely used, and the latest patterns can probably cope with most currently fielded threat systems, particularly man-portable SAMs. The supply of Soviet SA-7s to the Mujahadeen guerillas in Afghanistan has prompted the Soviet Union to make large-scale use of flares during air strikes against rebel positions. Flares are being expended at what has been described as "an awesome rate", a common Soviet tactic being for aircraft to attack in pairs, with one providing anti-missile protection while the other presses home the attack.

Shipboard systems

In addition to the aviation-related roles described above, chaff and flares can offer useful protection for

Below: An F-4 Phantom releases Space Ordnance Systems MJU-7/B infra-red flares over a simulated hostile environment.

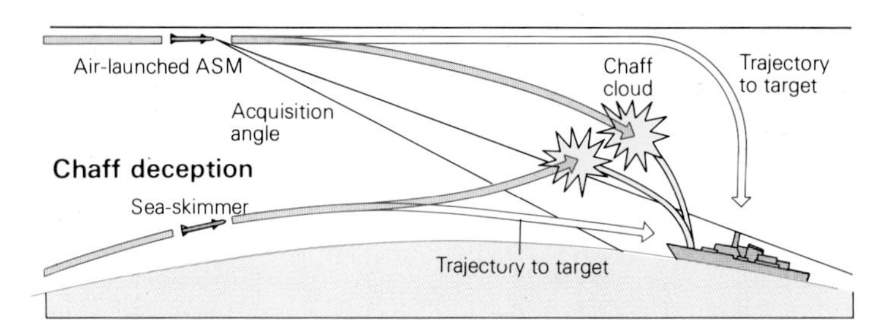

naval vessels, or even land targets such as AFVs. Despite the introduction of anti-missile weapon systems intended to engage and destroy anti-ship missiles, chaff still plays an important role in countering Exocet-type weapons.

For naval applications of this sort, rockets or mortars are used to carry the chaff projectile well away from the vessel it is protecting. Chaff shells may be fired from the vessel's main gun armament, and at times of high threat the ship's helicopter may also be used to deploy protective chaff clouds. During the Falklands War Royal Navy helicopters carried Chemring Chaff Hotel Broadband – a simple chaff package literally held together by a rubber band. Thrown by hand into the downwash from the rotor, this could create a frigate-sized target in only five seconds.

Probably the most combat-proven naval chaff system is the Vickers Corvus, which was extensively used by the warships of the "Operation Corporate" task force which liberated the Falklands. Troops landing on the British beachhead soon learned that the noise of Corvus rockets being fired was a sign that an air attack was only moments away.

The system consists of two trainable eight-barrel launchers for 3 inch rockets. These may be mounted anywhere on a ship's deck where there is sufficient space for the launcher to be loaded and trained. Corvus was originally designed to use the 3in Knebworth rocket, three of which could provide a multi-band reflective cloud, but the newer Plessey Broad-Band Chaff (BBC) rocket achieves the same effect with a single round. At a suitable point on the rocket's trajectory, the fuzing

Above: Distraction mode involves creating up to four chaff clouds around the ship at a range of up to 6,500ft (2km). These provide alternative targets to distract high- and low-level missiles.

Right: Each section of this Philips 9CM naval decoy dispenser holds 36 chaff cartridges.

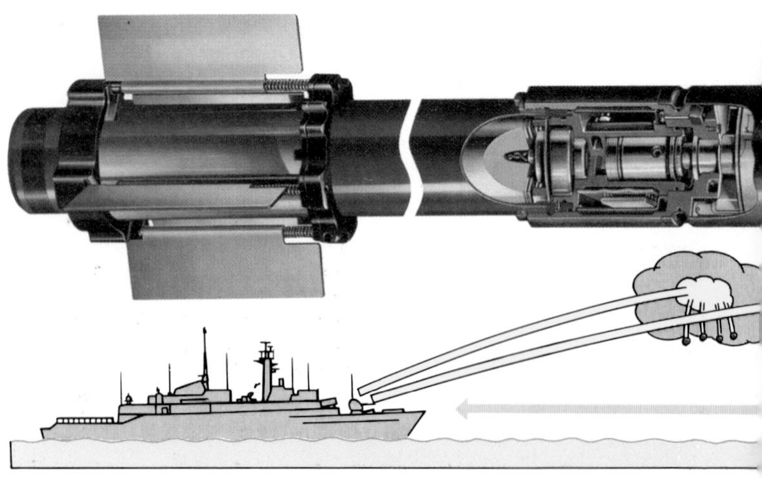

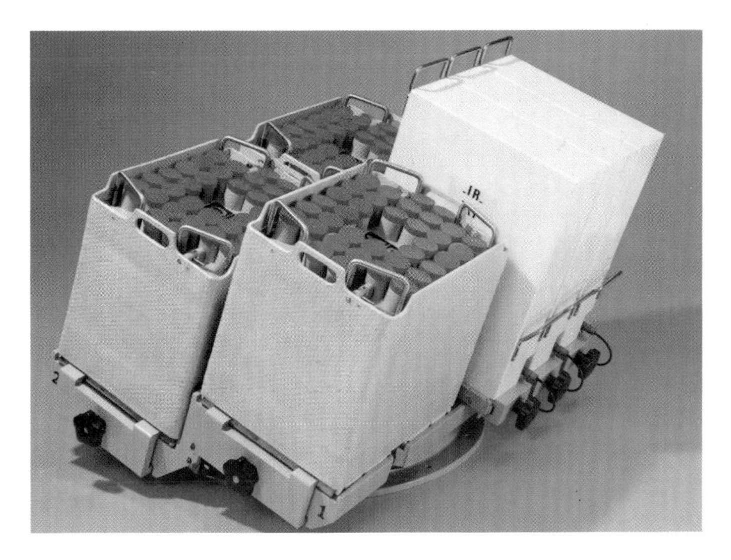

system separates the payload from the rocket motor, then dispenses the chaff payload in one of three modes described earlier.

Similar equipments are in service with most navies. Perhaps the most novel is the Saab EWS-900, with an eight-barrelled mortar energized by compressed air. Instead of using a small explosive charge to burst the chaff payload and disperse the contents, compressed air is used. This technique eliminates the hazards associated with pyrotechnic devices, according to the Swedish company.

Naval chaff can operate in several modes. If a radar-guided anti-ship missile is detected before its seeker has selected a target, a number of chaff clouds distributed around the vessel can cause the seeker to lock onto a false return signal. Alternatively, if use of the ship's ECM equipment causes a locked-on missile to break lock, chaff rockets may be used to present the missile with a false target on to which it can lock and subsequently home. Last-ditch defence of a ship against rounds detected at short range and in the final stages of flight may be attempted by deploying a chaff cloud within the range cell on which the missile seeker has locked, thus shifting the centroid of the apparent target.

Plessey Shield decoy system

Shield is based on an electronically-fuzed rocket (left) which can dispense chaff at short or medium ranges, or carry an IR decoy. When used in short-range seduction mode against an unidentified missile threat (below), a series of IR and chaff rockets are fired in order to move the radar and IR centroids away from the ship, causing a steadily increasing missile error.

AFV protection

Flares are likely to see use on future battlefields to protect AFVs from attack by IR-guided weapons. UK company Wallop offers the Guard system, which involves lightweight launchers on either side of the turret or hull of an AFV, each carrying two or three rounds of the company's 57mm rocket. Fired via an induction coupling, the rockets carry IR decoy payloads to an altitude of around 180ft (55m). The IR candle is ejected from the rocket then ignited, and while descending by parachute it radiates infra-red energy in the 3-5 and 8-14 micron bands. An alternative chaff payload produces a "bloom" of up to 5,400sq ft (500m^2).

Sophisticated decoys

These relatively simple decoys have more complex but highly classified counterparts. A combination of decoys and jamming can do much to confuse hostile surveillance radars, for example, so the US Navy is looking at methods of altering the radar signature of vessels in order to deny information on ships and task forces to Soviet radar. It is probably intended to confuse radar-equipped Soviet ocean reconnaissance spacecraft.

USAF B-52 bombers carried free-flying decoys from 1960 until the late 1970s. The ADM-20A Quail was a small pilotless aircraft weighing around 1,100lb (500kg). Released from the bomb bay of the parent aircraft, this unfolded its 66in (1.68m) span wings and flew for up to 250 miles (400km) under the power of a single J85 turbojet. During this 30 minute mission, the Quail used its built-in ECM equipment to simulate the radar signature of another B-52, creating a false target.

The follow-on AGM-86 SCAD (Subsonic Cruise Armed Decoy) project was soon upgraded by the addition of guidance and a warhead, finally flying as the AGM-86A ALCM-A (Air Launched Cruise Missile), ancestor of the ALCM-Bs which equip many SAC B-52s. No other bomber-launched decoy is known to exist, although hardpoints under the intakes of the Tupolev Tu-22M "Backfire" could be intended for such equipment.

Ballistic missile penaids

The highest classification is reserved for the penetration aids (penaids) carried by ICBMs and SLBMs to help them avoid ABM defences. The only system to have been publicly identified is the Mk-1A canister carried by the Minuteman II, but penaids are probably carried by all Western strategic missiles. No details of the techniques used are available, but a combination of chaff packages, active jamming, decoys and infra-red emitting aerosols seem probable.

The most likely initial Soviet response to the US Strategic Defence Initiative — apart from the start of a similar Russian programme — is the deployment of penaids and other decoys aboard Soviet strategic missiles on a massive scale. Decoys need not be heavy or physically bulky. If each re-entry vehicle carried by a missile were to be wrapped in a conductive balloon designed to be

inflated immediately after RV release, a much larger number of empty balloons could also be carried for release and inflation. To an ABM radar, these would seem indistinguishable from genuine RVs.

Since each balloon would weigh no more than 3½oz (100gm), decoys could outnumber genuine re-entry vehicles by up to a hundred to one, so the number of targets which any US defensive system would have to tackle could be numbered in hundreds of thousands. For this reason, the SDI programme is concentrating on methods of destroying missiles during the boost phase, before the process of RV and decoy release has begun.

EW smoke

"Make smoke" has long been a classic naval tactic. In the days when fleets of battleships and cruisers were the main instruments of sea power, the nimble destroyers would often be tasked with laying defensive smokescreens intended to mask the fleet from the approaching enemy. Smoke also finds application on the battlefield in vehicle-launched or hand-thrown form, or delivered by artillery and mortars. Were this the end of its applications, there would be no place for smoke in a book on electronic warfare, but recent developments are making smoke a valuable EW tool.

Conventional smoke is little problem to an army equipped with thermal imaging systems, since these can easily discern IR radiation through clouds which are opaque to visible light, but modern developments in smoke technology are

Above: Test launch of an ADM-20 Quail from a trials B-47. Quail equipped the B-52 force between 1960 and the late 1970s.

now making possible the creation of smokes which will blind IR systems. The British Army, whose Challenger and Chieftain tank units form an important part of NATO's front line in Germany, is leading the field in this new area of EW, and plans to fit its AFVs with anti-IR pyrotechnic systems. In addition to acting as infrared and optical smokescreens against viewing systems and laser designators, these should also offer protection from SACLOS (semi-automatic command to line-of-sight) anti-tank missiles such as TOW, Milan, HOT or the Soviet AT-4 "Spandrel", as well as fire-and-forget imaging infra-red homing missiles in the class of Maverick or Hellfire.

Two rival smoke systems have been developed in the UK. Schermuly's Multi-Band Screen operates by firing a vertical fountain of hot particles from a series of cartridges. These are designed to land outlet-side-up after launch from current patterns of AFV grenade launcher,

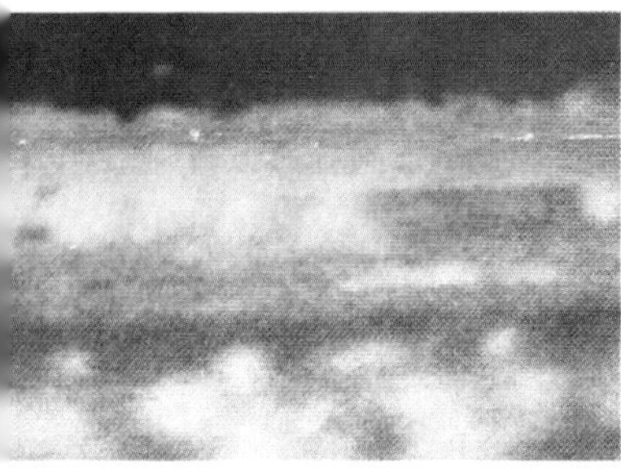

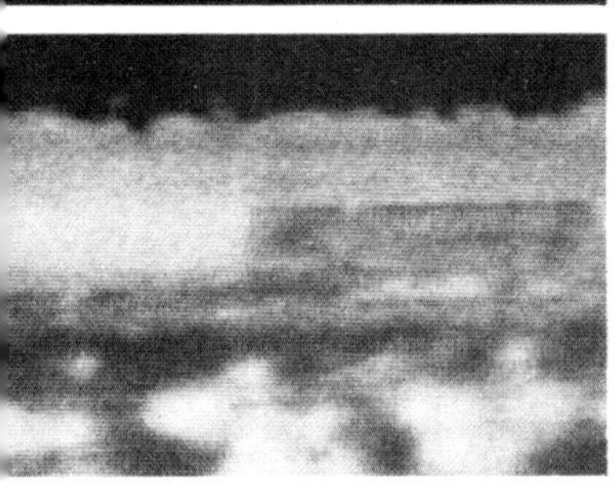

Above: Schermuly's multi-band screen provides cover against visual observation and infra-red sensors operating in the 3-5 and 8-14 micron bands. The left hand photographs show visual and IR targets (top), and a screen being fired from standard smoke dischargers of a Chieftain. (For security reasons the IR image is probably not that of a tank, but of a test vehicle). The right-hand photos show visual and IR views two (top) and three seconds later.

including the standard UK and US type and the West German 76mm system. According to the company, the current pattern of 12-shot launcher fitted to Chieftain could create a screen some 130ft (40m) wide at a cost of around £1,000 per firing.

Royal Ordnance Factory Glascoed has devised the VIRSS (Visual and Infra-Red Screening System) capable of providing screening over a 120° sector. Unlike the Schermuly system, this does not generate the screen from a source on the ground, but details of the method used to create the hot cloud are not likely to be divulged until the system is deployed. An experimental Chieftain fit features six launchers on either side of the vehicle, each comprising 20 individual launch tubes of small calibre.

The continuing battle

The development of countermeasures to decoys, and of improved decoys designed to defeat those countermeasures, is a continuous process. The best countermeasure to chaff is the use of CW or pulse-Doppler radar, but designers are already looking at the possibility of advanced chaff intended to be more effective against this type of equipment. The prospect of chaff being able to mimic the Doppler shift of a genuine target seems remote, but not impossible, while another future development might be the development of a chaff-like substance which instead of reflecting radar energy will absorb it. The prospect of optical chaff has also been discussed.

In the IR field, several new developments are aimed at devising IR decoys able to fool next-generation seekers. Possibilities here include aerodynamically-stabilized flares, and the use of pyrophoric materials which would spontaneously ignite when dispensed into a fine mist.

The future of expendable systems seems secure. Perhaps the best summary of their virtues is a comment made by Gowrishankar Sundaram, of the *International Defense Review*, a noted authority on EW. "It is these simplest of EW systems which have consistently delivered what they advertised – in the heat of battle".

Active Jamming

OBVIOUS targets in electronic warfare include enemy surveillance, target acquisition and tracking systems, plus the guidance systems of missiles and smart weapons. EW may also be used to good effect against speech, data and missile-guidance communications links.

Noise jamming

Jamming systems for use against radars operate in noise or deception modes. Noise jamming was the first technique to be used: dating from the infancy of EW, it is essentially a brute force technique, aiming to swamp the receiver of the target system with unwanted radio-frequency noise, so that the genuine and wanted signal cannot be distinguished from the background. If the exact operating frequency of the target is known, the jammer may be tuned to the same frequency, concentrating its power against the enemy.

The hostile radar may be pouring out kilowatts of signal in the direction of the target, but the return signal which it is attempting to detect is much weaker. Not only has the signal's journey out to the target and back again imposed its attenuation penalty, but much energy is also lost in the reflection process. Given the weakness of the return signal, it is relatively easy for the jammer to swamp the genuine echoes in noise.

Used against radars with PPI (plan position indicator) displays, noise jammers can "white out" a segment of the display by causing the radar screen to illuminate to its full brightness. Early radars with relatively broad-beam antennas were very vulnerable to this type of jamming, which could white out a large sector of the area being monitored. Modern radars tend to have such narrow beamwidths that only a tiny sector is affected, the resulting thin wedge-shaped sector of noise being referred to as a "strobe".

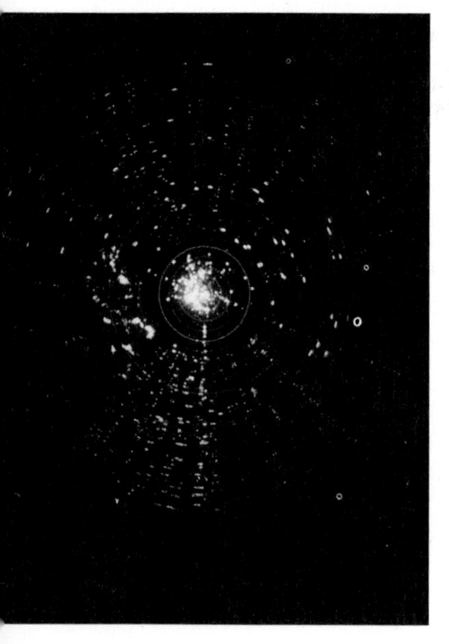

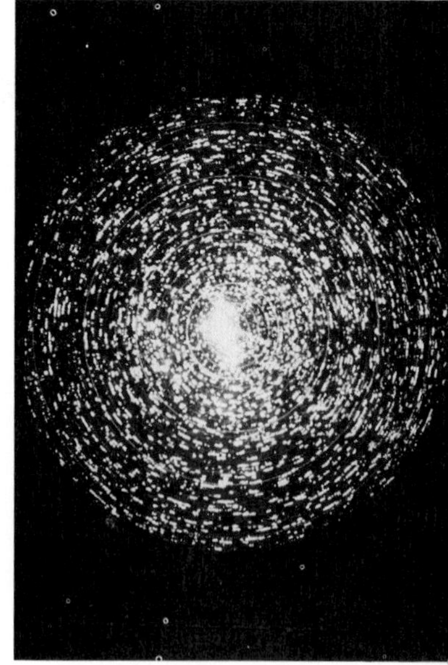

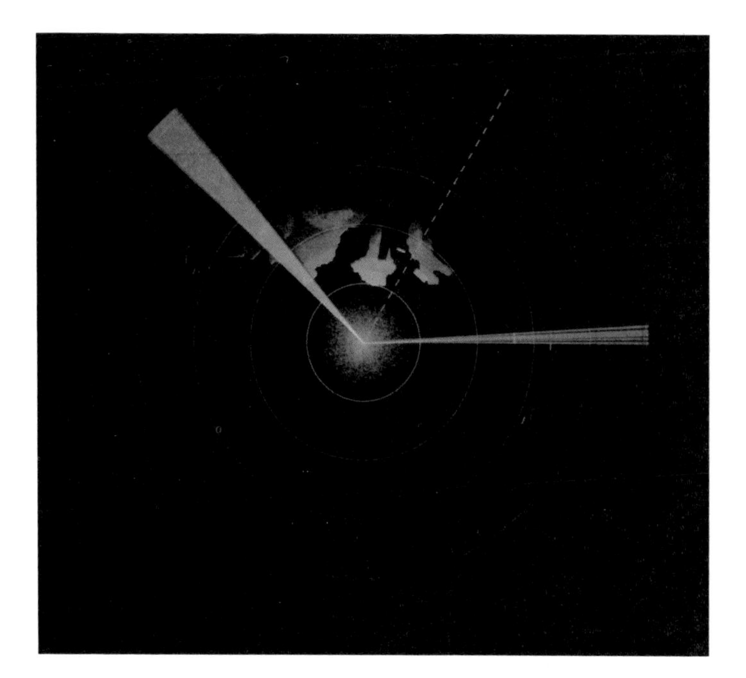

Above: Noise jammers can only disrupt a radar when the antenna of the latter is pointing towards the jammer. The resulting radial "strobes" are demonstrated here by a Ferranti radar simulator.

Spot jamming

Some noise jammers are set to transmit on a single frequency occupied by the hostile radar, a technique known as spot jamming. Although used effectively during World War II, spot jamming was countered in the 1950s by variable-frequency radars whose ability to switch to an alternative frequency enabled them to avoid the jamming signal.

This in turn involved one of two possible countermeasures. If spot jamming was to continue, the jamming platform needed to carry a search receiver and operator, so that the radar could be monitored and the jammer quickly retuned to the new frequency. Among the crew of a large bomber it was relatively easy to include an EW operator, so the 1950s saw the introduction in the USAF of three-man bomber crews, aircraft such as the B-47 Stratojet and B-58 Hustler carrying a pilot, a navigator/bombardier and an electronics operator. For a fighter or an attack aircraft, though, such manpower largesse was out of the question, so other techniques were adopted.

Left: A 360° radar display before and after being jammed by the EF-111A's ALQ-99E deception jamming suite. In the latter case (right), the entire screen is disrupted.

Barrage jamming

Barrage jamming spreads the output of the jammer over a range of frequencies likely to contain the threat – a technique which requires little elint effort since only the approximate operating frequency of the radar to be countered need be known. Its main disadvantage is that the jammer output is spread out along the spectrum instead of being concentrated on the actual operating frequency, so in a one-against-one engagement most of the energy is wasted. An alternative is swept-spot jamming in which the jammer operating frequency is scanned through a band of frequencies, disrupting each in turn. This will hit the hostile radar with the full power of spot jamming, but only at inter-

Above: The QRC-335 noise jammer seen here on an F-4 was an early US response to North Vietnamese defences. This design led to the later ALQ-101 and -119.

Right: The ALQ-119 seen here beneath the wing of a 354th TWF A-10A shows the dorsal gondola added to the pod in order to house new EW equipment added during upgrade programmes.

vals. Between bursts of interference, the radar may get enough data to maintain a track on its target.

Another anti-spot jamming system devised in the 1950s was frequency diversity. More subtle than simple frequency switching, this involved transmitting radar pulses on two or three discrete frequencies in turn. Simple spot jamming could not cope with this trick, so the EW designer was forced in the short term to resort to barrage jamming.

With the development of easily-retunable transmitters in the late 1950s, the radar was given even greater ability to switch frequency, creating greater problems for noise jamming operations. The old technique of manually searching for signals and retuning the jammer could be replaced by using the same transmitter technology to create electronically tunable jammers which could respond rapidly to changes in threat behaviour. This no longer had to be done manually, but could be achieved by automatic circuitry, eliminating the need for an EW operator.

Tactical jamming pods

In the early 1960s the USAF favoured noise jamming, having used the technique in its strategic bomber fleet. During the early stages of the Vietnam War, USAF tactical aircraft were protected against E- and F-band threats by the ALQ-71 noise-jamming pod, while the similar ALQ-72 covered I-band threats. The pods were originally ordered into development in the late 1950s as protection for F-100 Super Sabres and F-101 Voodoos, and were sufficient to deal with the most obvious threats — the E/F-band "Fan Song B" radar of the SA-2 "Guideline" and the I-band "Low Blow" set used with the later SA-3 "Goa".

Pod mounting had originally been tested by the US Navy with the early ALQ-31, but this large and cumbersome equipment was never accepted for general use. By the time the ALQ-71 and -72 were designed, smaller voltage-tuned magnetrons allowed jamming equipment to be packed into a 10in (25.4cm) diameter cylinder.

Noise jamming may be deployed in subtle ways by the addition of modulation to the basic signal. If frequency modulation is applied to a noise signal for example, the effect of the signal will be spread out over a range of frequencies. This can be particularly effective against surveil-

lance radars, covering the PPI display with many random spots of light.

The addition of modulation facilities to the ALQ-71 probably extended the life of this pod, but the follow-on ALQ-87 barrage jammer, which used broadband backward-wave oscillator (BWO) transmitter tubes was soon to follow. This started life as a G- and H-band system, with extra equipment for I-and J bands being added later. The system was modular, so could be rapidly reconfigured to cope with different threats (such as land-based radars and SAMs, or the radars carried by Soviet interceptors). The Westinghouse ALQ-101 was the first US jammer to enter production using TWTs, but it retained the traditional noise modulation.

Noise limitations

Noise jamming may be simple to implement, since very little needs to be known about the characteristics of the threats, but its effectiveness against sophisticated opposition is minimal. It also proclaims to the hostile radar that ECM techniques are being used. During World War II, German ECM operators gradually subjected the British radars guarding the English Channel to noise jamming. Day by day, the level of the jamming was increased, until the radars were sufficiently degraded to allow the German Navy to send two battlecruisers through the Straits of Dover. Radar was in its infancy in those days, and such tactics would hardly work today, when even the most poorly trained radar operator should be able to detect the presence of noise jamming.

By the closing stages of the Vietnam War, the limitations of noise jamming were becoming apparent, and a study of recent combat experience persuaded the USAF to make the switch to dual-mode jammers which could operate in either noise or deception modes.

Deception Jamming

The goal of deception jamming is to provide the hostile radar with false data. In its simplest form, the deception involves receiving the signal from the radar, processing it in some way, then retransmitting it in an attempt to persuade the radar to accept the spurious signal and derive false range or bearing information from it.

The USAF may have favoured noise jamming, but the US Navy has long preferred the more subtle deceptive approach, and the first US deception jammer was the US Navy's ALQ-19 angle-deception jammer developed in the late 1950s. This was soon followed by the ALQ-41 and -51, which both offered angle and range deception techniques. First USN aircraft to carry a deception jammer was the A-4 Skyhawk, which was fitted with the internally-mounted ALQ-51 system.

The USAF was not blind to the virtues of deception, having attempted to develop the ALQ-27 deception jammer for the B-58. This unit was unsuccessful, and probably overstretched the technology of the time, but the USAF returned to the deception technique when developing the Sanders ALQ-94 deception ECM set for the F-111, a unit designed for use against a specific class of threat systems. Not until the Westinghouse ALQ-119 – the first result of the dual-mode jammer policy – did the service turn to deception for general purpose jamming, and even then the -119 offered only limited dual-mode operation.

The first jamming pod able to work equally well in noise and jamming modes is the current Westinghouse ALQ-131, a 570lb (260kg) modular pod-mounted system able to cope with a wide range of threats, particularly the radars and guidance systems of air-defence systems. By selecting internal modules, the user can configure the pod to handle threats spread over one to five frequency bands, and modules are available to cope with all frequencies used by current anti-aircraft missile systems.

The TWT may have revolutionized radar design, but it was also a boon to the EW designer. For the first time, designers of jammers had the chance to create transmitters which could cope with rapid frequency changes and complex modulation forms, recreating exact replicas of the threat radar's own signals. By manipulating these, EW designers were able to make large-scale use of deception jamming techniques. Many have been devised, but only a few can be described here.

The simplest is to capture the individual radar pulses from the threat radar, and to transmit a series of replicas in rapid succession and separated by brief intervals of time before the next threat pulse arrives. Instead of a single return pulse, the radar now sees a large series which it interprets as a series of targets, all on the same bearing as the true target and indistinguishable from it – and the more rapidly the false pulses are transmitted, the greater the number which will be seen. If the radar receiver maintains a near-continuous watch for echoes (switching off only while the transmitter is starting the

ECM installations

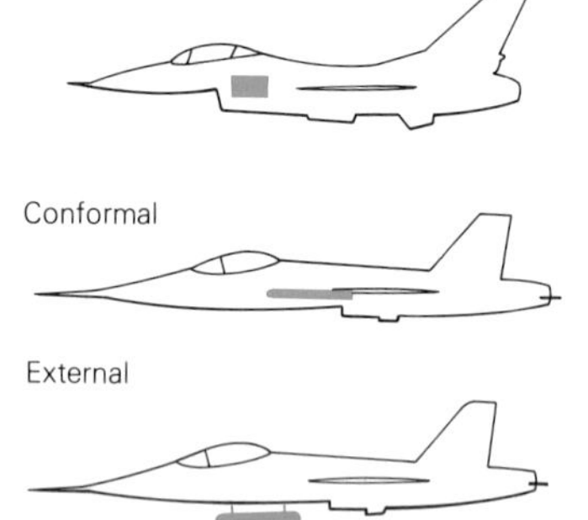

Above: Individual air arms favour different methods of carrying EW equipment. For a long time the choice lay between drag inducing external pods (which are easy to fit but can be masked by aircraft structure) and internal systems, but conformal systems are a recent compromise.

Above: Northrop ALQ-171(V) on a Swiss Air Force Mirage IIIS. A conformal version suitable for the F-5E is also available.

Right: The roll-stabilized antennas used by the Hughes SLQ-17 naval deception jammer.

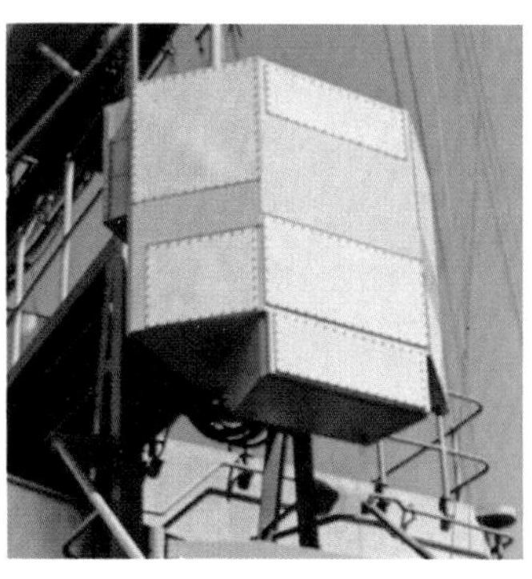

pulse on its journey) and the jammer transmits a long enough pulse series, the resulting false targets could stretch from the minimum to the maximum range of the radar.

Conformal pallet ECM

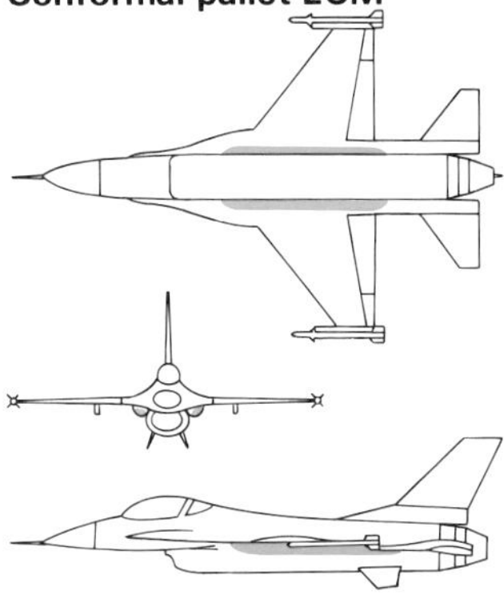

Above: Conformal fairings offer many of the best features of pod and internal fits. Mounted flush with the fuselage sides, they do not create a significant amount of drag, and leave all hardpoints free for fuel tanks or ordnance, but may be easily removed for repair or upgrading.

Internal ECM

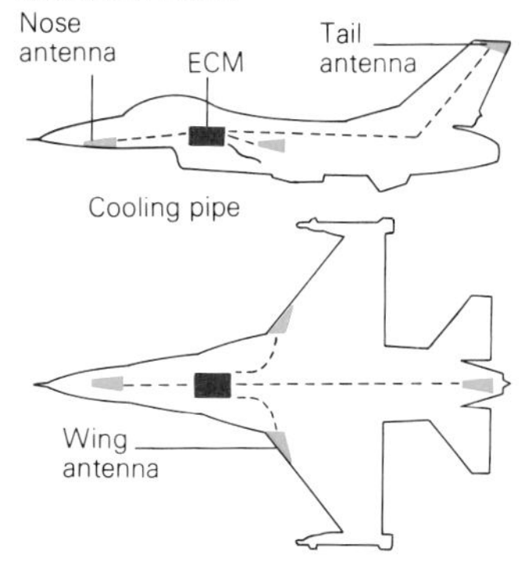

Above: Internal suites allow the antennas to be located in the best locations for all-round coverage, but the losses which result from passing signals through long cables or wave-guides are undesirable. Avionics space is always in short supply, making upgrades difficult.

Range gate stealing

This is not always the case. All the time the receiver is listening, electronic noise is being created – noise due either to its internal electronics or to external natural or man-made interference. Target-tracking radars get around this problem by only listening around the time that the echo is expected – a procedure known as "gating". Once a target has been selected, this is positioned in the centre of a range gate whose width is sufficient to allow for the effects of changes in range resulting from target velocity. With each successive pulse, the position of the range gate is readjusted to keep it centred around the target. The stage is thus set for a deception technique mentioned earlier in this book – "range gate stealing". Also known as "track breaking," this can have a devastating effect on radar trackers.

By retransmitting a replica of the true echo at the same time and with no time delay, the jammer simply increases the apparent strength of the target echo. As we have seen, radar returns can fluctuate in amplitude depending on target attitude, so this increase in amplitude does not seem unduly abnormal. The receiver's automatic gain control (AGC) circuitry simply adjusts the gain to match the new apparent signal strength. The radar and the circuitry are now responding, not to the true echo, but to the fake being transmitted by the jammer.

When the jammer starts to introduce a tiny amount of delay before returning the false pulse, the radar's automatic circuitry used to maintain the position of the range gate will adjust its position to re-centre the gate around the jammer's relatively powerful fake pulse rather than the weaker original. The gate has thus been "stolen", and its future location will be determined not by the true target but by the jammer. By steadily increasing the time delay, the jammer moves the gate further and further away from the true range, giving the radar false range information which will induce an error into gun or missile-control commands. By turning itself off, the jammer can leave the radar with an empty gate and apparently vanished target. Once the true target has been re-located and the gate repositioned, the jammer can simply steal it anew.

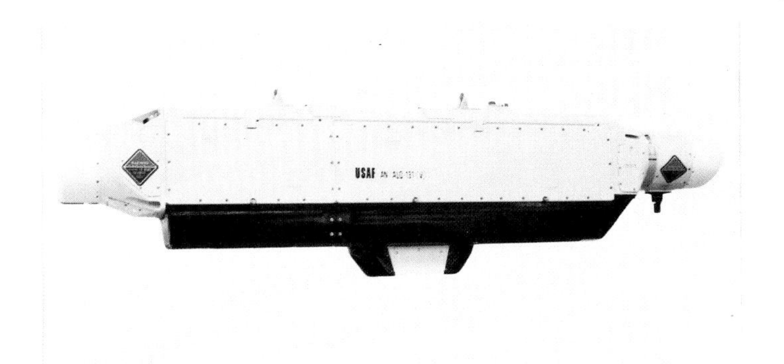

Above: Jamming involves high power levels — red and black markings on this ALQ-131(V) warn ground crew to stay at least 15ft (5m) away while the pod is carrying out test transmissions. The ALQ-131 can be configured to cope with any known SAM system by selecting different internal modules operating on the appropriate frequency.

Above left: A USAF F-16 of the 10th Tactical Fighter Squadron, 50th Tactical Fighter Wing, equipped with a Westinghouse ALQ-131 jamming pod.

Inverse gain jamming

Surveillance radars are vulnerable to another deception technique which can be used to create errors in bearing. As the radar beam's circular scan pattern carries it onto and past the target, the jammer (given a suitable receiver) will be able to detect this. When the signal from the threat is at maximum, the radar beam is pointing directly at the jammer, and the true return echo will be at its maximum. The radar works on the rule that the true target bearing is that on which the strength of the observed return signal is greatest. This simple rule plays directly into the hands of the jammer designer.

Once again, he creates a jammer which will recreate a replica of the true echo. In this case, the fake echo must be many times stronger than the true echo, because it will be transmitted not when the antenna is pointing directly at the target but just before or just after, while the jammer is still within the beamwidth (effectively the field of view) of the threat radar. When the target is in such a position, it is still returning reflections to the radar, but since it is no longer on the beam centreline, these echoes are weak. The radar assumes that since these echoes are weak, they do not represent any true target bearing — the simple rule mentioned above.

By deliberately boosting these faint echoes by adding its own replica at the right moment, the jammer can persuade the radar's automatic circuitry that the target is now on the beam centreline — a centreline displaced by several degrees from the true target bearing. Since the power of the jammer must be adjusted to be inversely proportional to the strength of the signal received by the jammer from the radar's transmitter, this technique is known as "inverse gain" jamming.

Combined use of bearing and range deception techniques can generate large numbers of false targets on a radar screen — signals which maintain tracks, manoeuvre and behave in every way like genuine targets. Used ruthlessly, such techniques can completely overload the radar. During an unclassified and therefore strictly limited demonstration of the ALQ-99 jamming system carried by both the Grumman EA-6B Prowler and the EF-111A Raven, the author was shown how the PPI display of a surveillance radar could be totally whited out, not by noise jamming but by filling the screen with so many false targets that these merged into a continuous "snowstorm" of close-packed echoes.

Inverse gain deception jamming

Below: Deception jamming involves transmitting fake return signals which the victim will accept as genuine. These are often sent when the antenna of the victim is not directly facing the jammer, so they must be very powerful in order to "leak"

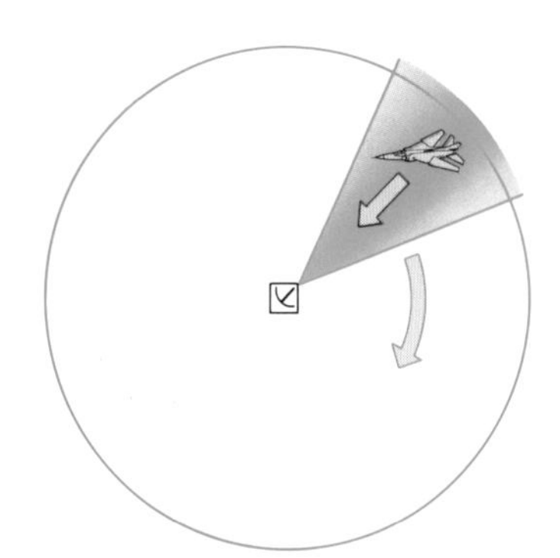

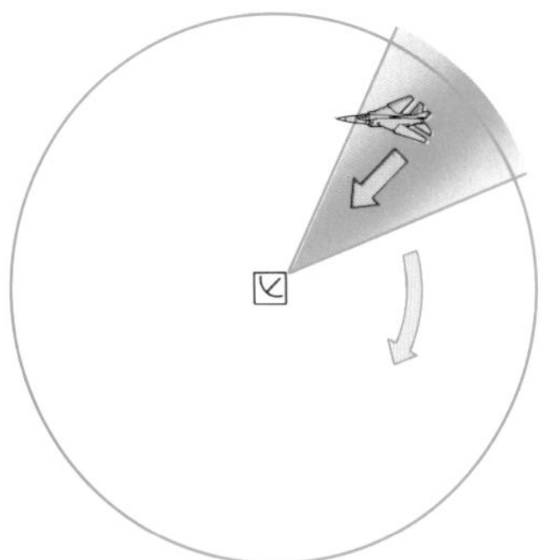

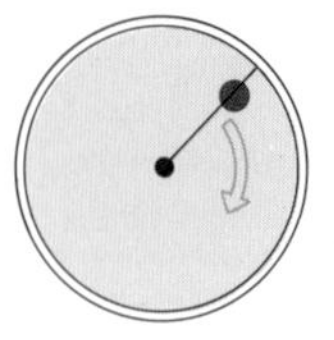

Normal operation: the antenna is pointing at the target aircraft, which returns a genuine echo seen on the PPI as a target.

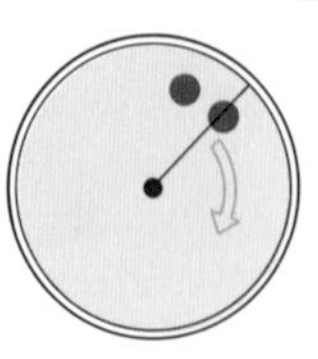

With the aircraft on the edge of the beam, the radar will reject the true target in favour of a powerful fake echo apparently on another bearing.

Left the antenna of the "Gun Dish" radar of the ZSU-23-4 Shilka anti-aircraft vehicle uses conical scanning, so is vulnerable to deception jamming.

into the antenna via a sidelobe. The fake pulse will then be of an appropriate strength to be accepted as genuine.

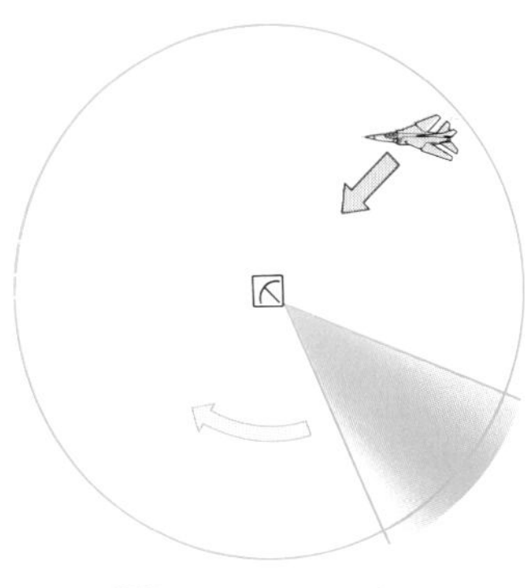

By transmitting a massive fake echo later in the scan pattern, the jammer can create another target on a totally false bearing.

Conical scan deception

Deception techniques similar to inverse gain jamming can have a disastrous effect on some types of tracking radar. Since a tracker must keep its antenna directly pointed at the target, it must use some technique to measure the presence and direction of any aiming error. The most common is conical scanning, a process of scanning a volume of space surrounding the target. This is normally done by off-centre rotation of the antenna feed or even the entire dish.

If the antenna centreline points directly at the target, the latter will be equally illuminated by the slightly off-centre beam at all points in the scan pattern. If an aiming error is present, on the other hand, the beam will return a stronger signal at that position in the scan, and the observed signal strength will vary at a frequency equal to that of the conical scanning motion. By noting the magnitude and timing of this variation, signal-processing electronics can determine the direction and magnitude of the aiming error, and a steering command can be generated to bring the target back on the centreline, at which point the signal will again be constant.

Conical scanning is widely used in systems such as the "Skip Spin" and "Spin Scan" radars carried by the Su-11 and MiG-21, the AIM-7E and AIM-7F versions of Sparrow and on the J-band "Gun Dish" radar of the ZSU-23-4 Shilka self-propelled anti-aircraft gun, but is very vulnerable to countermeasures. If the jammer designer knows the conical scanning speed, he can design his system to transmit a false signal modulated at that frequency. Once received by the victim, this will be indistinguishable from a genuine aiming error, causing a false steering command to be generated. As a result, the antenna will be deflected from the genuine target bearing and may even lose lock.

In the case of an active threat radar such as "Gun Dish", elint observation of the signal will determine the scanning rate, allowing the design of the jammer to be optimized, but the seeker on semi-active missiles such as Sparrow does not transmit. Targets are illuminated by the radar of the Sparrow-equipped fighter, and the missile homes on the energy reflected by the target. If scan rate information cannot be obtained via espionage or analysis of captured equipment, the jammer can only scan a band of likely frequencies, so will disrupt the seeker at intervals rather than constantly.

Monopulse and CW radars

There are several countermeasures to this jamming technique, and a clue to one is given by the example of Sparrow – use separate transmitting and receiving antennas, applying conical scanning only to the latter. This technique is known as LORO (Lobe On Receive Only) and is used by Soviet systems such as the SA-2 "Guideline".

A more effective method of eliminating the ECM weakness of conical scanning is that used by the latest AIM-7M version of Sparrow and many modern radar trackers, including the nose-mounted radars of modern jet fighters. This involves dividing the antenna beam into four parts, each slightly displaced from the antenna centreline – for example upper, lower, right and left parts. If the antenna is pointing directly at the target, all four beams will see target echoes of similar strength; if not, the aiming error can be deduced by comparing the strength of the echoes received from the four beams. A conically-scanned sensor must complete one revolution in order to measure target position, but the four-beam system can establish target position near-instantaneously by observing the echo from a single radar pulse. For this reason, radars of this type are referred to as "monopulse" radars. Like LORO radars, they are difficult targets for deception jamming.

Another difficult threat is con-

Below: During the 1973 Middle East War, Israeli pulse-radar jammers proved useless against the CW guided SA-6.

Above: Britain's Lightning fighter was the first to be fielded with a monopulse radar in the form of the Ferranti AI.23B.

tinuous-wave (CW) radar. Since this does not generate pulses, the pulse-manipulation techniques described earlier are ineffective.

One invariable rule of EW is that for every sensor technique, there is at least one suitable countermeasure, and often several. Similarly, for every countermeasure, there is often a counter-countermeasure, and given the correct jamming technique, monopulse, LORO and CW radars are all vulnerable to disruption.

One method is based on a phenomenon known as radar glint — the apparent fluctuation in the angular position of a radar target. A telescope accurately aligned with the antenna of a radar tracker would show that the centre of a moving target as seen by radar does not coincide with the physical centre of the target, but shows rapid short-term fluctuations. Strange though it may seem, the radar "target" may even briefly lie outside of the physical confines of the actual target — the result of the radar being confused by differences in the phase of the radar echo from different parts of the target.

Having now been exposed to the devious logic of deception jamming, the reader might be forgiven for wondering whether a radar could in some way be permanently "persuaded" that a target's indicated position did not coincide with its physical location. This is exactly the approach taken by EW designers attempting to jam monopulse or LORO trackers. If phase differences in reflections from different parts of the target cause the desired effect, then the trick is simply to make sure that they are present in abundance. As with previously described deception techniques, the genuine signal is received, doctored, then retransmitted. In this case, instead of introducing a timing error, a phase error is created.

Cross eye and buddy mode

One technique for doing this, known as "cross eye" jamming, involves transmitting the fake echo from two antennas spaced as widely apart as possible (for example, on the aircraft's wingtips) after introducing a deliberate phase difference. For best

results, this phase difference must be correctly matched to the bearing of the threat radar — easily done by using the same antennas and phase-shifting equipment to receive the hostile signal and adjusting the phase difference between the antennas until the two signals are opposite and equal so that they cancel each other out at the EW receiver.

This phase difference is then applied to the fake echoes, which can induce an error in the threat radar equal to several times the spacing of the antennas on the aircraft. This may be enough to cause radar-guided flak or missiles to miss their target and, if greater than around half the threat radar's antenna beamwidth, will cause the set to lose lock.

"Buddy mode" jamming is a variation of this technique. Two aircraft must fly in close formation within the threat radar's antenna beam, each

retransmitting fake echoes of approximately equal power. According to Soviet engineers, the error induced in the radar can be equivalent to half that of the distance between the two aircraft.

"Image buddy mode" even manages to dispense with the second aircraft, relying on the fact that an aircraft or missile flying at low level creates a radar reflection of itself in the terrain or water over which it is flying. At the very low altitudes used by sea-skimming missiles such as Exocet, missile and reflection may be so close together that they are both in the radar main beam. The radar therefore detects two targets in its field of view, one directly and the other via reflection in the water. This can introduce a severe case of confusion in the threat radar, which may end up wildly switching its aim from one to the other and back again at rapid intervals.

Above: A radar may dither between two targets flying in close formation. Buddy mode jamming creates the same effect.

In the case of a manned aircraft, the cruise height will be enough to ensure that the reflection does not lie in the main beam of the threat radar. It will still be seen by the radar, thanks to the existence of antenna sidelobes, but the echo power entering the antenna by this back-door route will be very small. The jammer's task is to give nature a helping hand by aiming towards the ground a massively powerful fake echo which, even after the losses caused by reflection and reception via the antenna sidelobes, will still rival the power of the genuine echo returned by reflection from the target. The result — as seen by the radar — will be indistinguishable from that of buddy mode jamming.

Velocity track breaking

"Velocity track breaking" is a technique similar to range gate stealing, and was devised to counter pulse-Doppler radars: widely used for look-down/shoot-down applications, radars of this type observe the Doppler shift of the return signals in order to determine target velocity. Instead of applying a slowly increasing time delay to fake return signals in order to steal the range gate, velocity track breaking involves applying a slowly increasing frequency shift which the victim will interpret as a change of target velocity. By stealing the velocity gate in this manner, the jammer can induce a growing error in target tracking and can even cause the radar to lose lock when the jammer is turned off.

Requirements and reservations

All deception techniques are designed to subvert part of a radar set's automatic circuitry, confusing electronics which have in the past had little built-in intelligence. This relatively subtle approach does require accurate data on the characteristics of the threat system, and also the techniques which the latter uses. If the deception technique is not properly matched to the threat, the jamming will be useless.

Combat experience in Vietnam also showed that deception jamming can be less effective against older patterns of radar, since such sets rely on the large degree of human monitoring and a trained operator may soon learn how to spot the effects of deception. Noise jamming is likely to prove a better solution in such circumstances.

Another problem with deception jammers is that they must process the pulses from every radar being jammed. If only a limited number of radars are in action this is no prob-

Above: Deceptive ECM is of limited effectiveness against older manually-operated radars.

Top right: Thomson-CSF DB3163 jammer on a Mirage F1.

Above right: The Swedish SATT AQ-800 pod is intended for training radar operators.

lem, but the high emitter densities displayed by Warsaw Pact defences could overload the jammer's capacity to deal with them.

The problem is similar to that experienced by radar-warning receivers. Once again the solution is to rely on automatic signal-identification and prioritization: the jammer needs to be able to compare the parameters of the signals detected by its built-in receiver with threat data stored in its electronic memory; once those representing the greatest threat to the aircraft have been identified, the jammer assigns the power available from its various jamming transmitters accordingly.

This technique is known as power management. The type of jamming to be used will be chosen to match the threat, and in the most advanced systems the transmitters will be turned off momentarily at regular intervals, allowing the receiver to monitor the hostile transmission and assess the effect of the jamming. If the latter is not proving effective, a different jamming method may be automatically selected.

Non-US jammers

Since most of the jamming systems described so far have been of US origin, it must be stressed that these were chosen only because the United States is more open in discussing EW than other nations. The Soviet Union pays great attention to jamming. For example, underwing payloads for the MiG-21 include an ECM pod, while the MiG-21R reconnaissance variant carried ECM in wingtip pods; the Su-24 "Fencer" is probably the first Soviet tactical fighter to carry built-in jamming systems, but bombers such as the Tu-16 "Badger", Tu-95/142 "Bear", Tu-22 "Blinder" and Tu-22M "Backfire" all carry internally-mounted ECM equipment. Unfortunately, no details are available.

Jamming equipment is also offered by West European companies such as Marconi (UK), Thomson-CSF (France), Selenia (Italy), Ericsson (Sweden), and AEG-Telefunken (West Germany). Israel is a relative newcomer to the field, but several companies including Elta now offer jamming equipment of their own design.

IR, EO and Sonar Jamming

Infra-red (IR) seekers and trackers are also vulnerable to deception countermeasures. The rotating-scan operating principle on which most non-imaging IR guided weapons operate may be simple and cheap, but the use of a regularly rotating reticle is a weakness exploited by designers of countermeasures. The basic method is similar to that used to counter conically-scanned radars: deliberate generation of a false source of flickering IR energy which the missile will interpret as evidence that the seeker is not pointed directly at the target. This false error results in spurious steering commands being sent to the seeker head, driving it away from the target bearing.

Typical US IRCM systems include the Loral ALQ-123 (pod-mounted on the A-6 and A-7) and the Northrop AAQ-4 (internally mounted) and AAQ-8 (pod-mounted) systems. Such systems are sometimes hastily deployed as a result of combat experience: during the Vietnam War AAQ-4 IR jammers were mounted within the main cabins of Lockheed C-5A Galaxy transports as a defence against man-portable SA-7 "Grail" anti-aircraft missiles. The big transports took off from Vietnamese airfields with the tail ramp lowered and AAQ-4 running, so that any SA-7s fired from positions near the field would be jammed as they closed in on their pursuit-course attack.

Many IR countermeasures systems such as the Sanders ALQ-144 and -148 (pod-mounted on helicopters and OV-1D FAC aircraft respectively) use a continuously operating source of radiation, and rely on a mechanical shutter system to provide the necessary modulation to confuse seeker systems. Mechanical modulation may be simple to design but results in an inflexible equipment, as the modulation pattern is fixed by the configuration and rotational speed of the mechanism, and cannot be changed easily to cope with modified threats. Mechanical modulation also lacks the sophistication needed to deal with more advanced threats such as Soviet ground-based IR trackers.

The alternative adopted by the AAQ-4 and -8 (and probably used on the -123) is an electronically modulated IR source such as an IR-emitting lamp, though the rapid on/off switching dictated by modulation patterns reduces the life of

Below: Sub-units of the Northrop MIRTS modularized IR jammer.

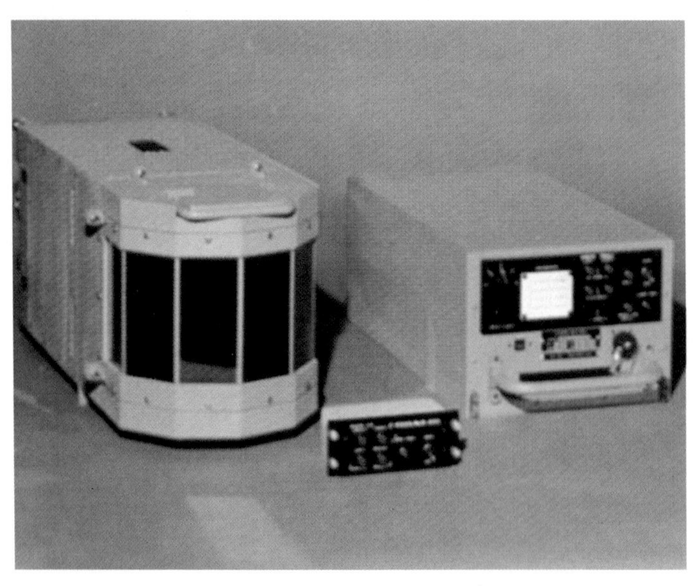

these components, so the jammer cannot be run continuously, and a warning receiver must be used to indicate that an IR threat is present.

Several new US IR countermeasures projects have been identified, including the Advanced IRCM programme. These are under development to meet the threat posed by the latest generation of IR seekers, trackers and FLIR systems. Reports that a new guidance system resistant to IR countermeasures is being developed for the Chaparral IR-homing point-defence SAM are evidence of the growing deployment of IR countermeasures by the Soviet Air Force, while US Navy plans to improve aircraft decoy techniques suggest that similar improvements are expected in Soviet IR-guided SAMS.

EO jamming

Visually or electro-optically guided surface-to-air and air-to-air weapons may also be attacked by EW. Detection systems can warn aircrew that they are under attack, while decoys and jammers may be used to confuse tracking and homing devices. Most of this work is classified, but it is possible to identify a number of programmes.

Above: Northrop's pod-mounted AAQ-8 infra-red jammer uses electronic rather than mechanical modulation, so may be quickly given new modulation patterns in order to cope with newly identified threats.

Below: The Xerox ALQ-157 IR jammer, seen here being installed on a Chinook helicopter, is designed to confuse heat-seeking missiles. Two transmitters are carried, one on each side.

An advanced EOCM pod, originally codenamed Compass Hammer, is being developed by both the USAF and USN. Flight trials of experimental pods started in the early 1980s, and the Coronet Prince programme now in advanced development applies technology evolved under Compass Hammer to an operational system intended to enter production in 1990. Coronet Prince is usually described as being an optical countermeasures pod, but has also been described as an "optical threat acquisition and cueing system". Working under Compass Hammer, Westinghouse and Martin Marietta developed rival systems carrying the unofficial designations ALQ-179 and ALQ-180, one of which could be adopted in modified form by the Coronet Prince programme.

The Expendable Laser Jammer (ELJ) is about to commence flight trials, while development of at least three other EOCM systems has been reported. One is for an air-to-air application, another is a counter to projected electro-optical threats, and a third is intended to cope with laser-guided weapons.

Above: The NKC-135 laser test-bed has burned out the seekers of AIM-9 Sidewinder missiles.

Right: Weapon aimers could be dazzled or blinded by lasers.

Sonar countermeasures

Like radar, IR and electro-optical systems, sonar may be attacked by EW. Since World War II, surface ships have towed acoustic decoys intended to lure homing torpedoes, and during the Falklands War, one Royal Navy ship is reported to have lost its towed decoy when the latter was attacked by a friendly ASW torpedo.

The US Navy's Project F34-371 is intended to develop improved countermeasures to protect surface ships from torpedo attack, and decoys which will allow submarines to evade hostile ASW vessels equipped with new patterns of Soviet sonobuoys whose deployment has been predicted by US Intelligence. Advanced technology is also being developed to counter perceived future sound-navigation and sonar systems, and also submarine-launched underwater weapons.

USAF
ASD
53123

Communications Jamming

Radio links are another prime target for EW, and one which has been exploited in several recent conflicts. Voice, data, and even missile-command links are all vulnerable to jamming. If their operation is disrupted, the result can be chaos which an enemy will be quick to exploit.

During the Beka'a Valley air battles which followed the Israeli invasion of the Lebanon, IDF-AF aircraft successfully jammed Syrian surface-to-air communications links, depriving Syrian Air Force fighters of the ground control on which their Soviet-style operating tactics were based. The resulting near-paralysis of the Syrian defences may explain the 80 to 2 score which the Israelis claimed in air-to-air combat during the first week of the campaign.

A cruder example of "comjam" appeared during the 1982 Falklands War. The British electronic order of battle in that conflict contained several unofficial and improvised EW systems, the most bizarre of which was probably a makeshift communications jammer based on a faulty electric motor.

Communications jammers normally act as noise generators, with the output of the jammer transmitter either being spread over a range of frequencies used by the enemy, or selectively focused on a few key transmissions. Once again, spot jamming is the most effective, since it concentrates the effect of the jamming, while leaving most of the frequency band unjammed, and thus free for friendly communications.

The efficiency of wideband jammers is reduced by the ratio of the total bandwidth to effective bandwidth, and the equipment makes no distinction between friendly communications and those of the enemy, disrupting both. Equipment of this type is generally considered a secondary threat.

In order to attack one or more transmissions selectively, the jammer or its operator must either predict the frequency at which the target transmission will occur at a given moment or monitor a range of frequencies to determine whether transmission is being made.

Frequency-hopping radio

Jamming of frequency-hopping radio is virtually impossible given the current state of technology. A hostile elint receiver may be fast enough to detect the short burst of signal on a given frequency which is transmitted before the radio it is trying to intercept hops to another frequency, but this achieves little. The elint

Below: The US Army's Piranha tactical jammer in action at a front-line position.

receiver has no way of identifying to which of the many frequency-hopping systems active at any one time this tiny burst of signal belongs. The only electronic systems which know the frequencies on which subsequent pulses will be transmitted are the radios themselves. The elint system is faced with a large number of unidentifiable pulses. If a larger number of frequency hoppers are active – the likely case in a combat situation – the signal-discrimination facilities of the EW system will become saturated and its efficiency considerably reduced.

Frequency-hopping equipment therefore becomes more effective as a greater number of systems are deployed.

For the moment, frequency-hopping radio links seem secure, a Western view apparently shared by the Soviet Union. According to one report, the Soviet government uses frequency-hopping radio links for highly classified communications. For maximum security a relatively fast hop rate is used – more than 1,000 per second.

Future systems

Obsolescence is a problem with any jamming system, and particularly with deception systems. In the past, hardware was often specifically built to deal with identified threats. A new threat could involve modifications to existing hardware, or even completely new sub-systems. For this reason the design of such last-generation systems as the USAF Westinghouse ALQ-131 was modular in order to facilitate upgrading.

Newer equipment such as the USAF/USN Westinghouse ALQ-165

Left: An Israeli IAI Elta EL/K-700 VHF communications jamming station mounted in an M113 APC. Designed for use at brigade or divisional level, it uses a directional Yagi antenna.

Below: This VHF jammer consists of two manpack radios, control unit and power amplifier.

Left: The USA takes the EW threat to its orbiting spacecraft seriously. With the launch of the Tracking and Data Relay Satellite (seen here being cleaned to remove accidental pre-launch contamination) came evidence of how anti-EW features are being incorporated into US designs. When the spacecraft ran into technical difficulties with its control system soon after launch in early 1985, NASA was reluctant to release detailed information. To do so might compromise encyrption facilities built into the command and control system as a guard against unauthorized attempts to send spurious commands, Agency spokesmen explained.

Right: A pre-launch antenna erection test of the Tracking and Data Relay Satellite. Impressive though this orbiting "dish" antenna may be, it would be dwarfed by the much larger folding antenna carried by the Rhyolite sigint spacecraft.

Advanced Self-Protection Jammer (ASPJ) or the Marconi Sky Shadow are software-controlled, using built-in computers to match countermeasure techniques to threats in accordance with a combination of established threat data and real-time information obtained from the system's built-in receiver/signal processor.

Future jamming systems will have to cover a much wider spectrum than current types. A tantalizing glimpse of the shape of jammers to come is given by the joint USAF/US Navy INEWS which will probably be fitted on next-generation US aircraft such as the Advanced Tactical Fighter (ATF) and the Northrop Advanced Technology Bomber (ATB).

Currently under study by several US industrial teams, and due to enter full-scale development in 1988 or 1989, INEWS will be a modular system covering not only the conventional radar bands, but also millimetre-wave, infra-red and laser frequencies. It will probably be highly integrated into the avionics system of the parent aircraft, using a digital databus and sharing the radomes or even the actual antennas of other equipments. Antennas can act as good collectors and re-radiators of radar energy, increasing the aircraft's effective target size. This is highly undesirable in a "stealth" aircraft, so such sharing of antennas or radomes will be a significant development in the struggle to minimize radar cross section.

Research and testing

In order to assess the vulnerability of weapon systems to ECM (and to sigint), and to devise suitable counter-countermeasures, equipment must be thoroughly tested. The US Army, for example, maintains programmes which monitor the EW vulnerability of surface-to-surface weapons, air-defence missile, C^3, tracking and surveillance radars and night-vision systems. These assess current vulnerabilities, the threat posed by known (and even nominally friendly) EW systems, and study suitable ECCM counters. Recent studies have covered Pershing II, Lance, TOW, Hellfire, Copperhead,

Improved Hawk, Stinger, US Roland, Patriot, the Sergeant York self-propelled AA gun and TPQ-36/37 radars. This work often results in technology which may be applied to new systems: the frequency-hopping system used in SINCGARS was originally developed under such ECCM studies.

New applications for jamming are emerging. It is reasonable to assume that Soviet missile engineers are working on fire-and-forget weapons able to detect targets after launch and carry out an autonomous discrimination and lock-on process. The fact that the USAF has abandoned the Hughes Wasp missile of this type is hardly likely to restrain Soviet design efforts. Work is now under way in the United States to devise technologies able to confuse and defeat such systems.

As sophisticated command and control systems, highly co-ordinated EW systems, and multi-mode seekers and trackers enter service, suitable countermeasures must be devised, and work is already under way in the US – and presumably in Western Europe and the Soviet Union – to develop the technology needed to deal with such threats.

Jamming in space

Space is another future arena for EW. Advanced technologies which will reduce the susceptibility of US satellite systems to jamming and physical destruction are being explored, and anti-jamming facilities have already been built into some US spacecraft. Problems early in 1985 with the Tracking and Data Relay Satellite led to an admission by NASA that the spacecraft's control system incorporated encryption facilities to protect it from interfering signals.

Although the Space Shuttle was developed by NASA, it will be used as a carrier for US military spacecraft, and could therefore be threatened by Soviet ASAT systems. The US Navy is reported to have carried out Shuttle-related experiments as part of its wide-ranging Satellite Communication and Defense programme, and to have informed NASA of the results.

Anti-radar Weapons and Aircraft

THE ultimate anti-radar weapon is a 1,000lb bomb applied to the offending radar equipment. A jammed radar and its crew survive to fight another day, but a destroyed radar and a dead or injured crew are out of the war for good. The US armed forces are firm believers in this philosophy, and within days of the first loss of a US aircraft to SA-2 fire during the early stages of the Vietnam War, the offending missile sites were being pounded into silence by air strikes.

Missiles and aircraft custom-built for anti-radar strikes, and able to home on hostile radar transmissions, are a US speciality. These are largely intended to counter land-based threat radars used for the control and guidance of surface-to-air missiles, but can also be used against naval radars. However, attempts to develop air-to-air anti-radar missiles were abandoned, as will be recounted below.

First of the USAF's anti-radar aircraft was the F-100F, a rebuild of the two-seat Super Sabre. Designated "Wild Weasels" and fitted with the APR-25 radar-warning receiver and the 2-4GHz IR-133 panoramic receiver, these were rushed out to the 388th TFW at Korat RTAFB in Thailand from November 1965 onward. Operational from the following month, the Weasels made their first attack on an SA-2 site in April 1966. Only a small number were built, since the aircraft lacked the performance and more sophisticated avionics of the follow-on F-105F, and the only major improvement introduced during its combat career was the in-

stallation of the AGM-45 Shrike anti-radar missile in 1966.

AGM-45 Shrike

The end product of a development programme started back in 1962, Shrike was based on the AIM-7 Sparrow, but used a passive homing radar seeker designed to steer the weapon down the beam of threat radars. Shrike was initially used in the less defended areas of North Vietnam until operational tactics had been proven, but by the summer of 1966 were ranging deep into hostile airspace,

Above: Compared with the earlier Shrike, the AGM-78 Standard anti-radiation missile has a heavier warhead and substantially longer range. The dark panels aft of the nose are part of the fuzing system.

Below: This F-105F Wild Weasel version of the Thunderchief in action over Vietnam carries AGM-45 Shrike under the outer wings. Shrike's relationship to the AIM-7 Sparrow from which it was developed is obvious.

attacking SAM sites and destroying the specialized radars such as "Fan Song" which were used for missile tracking and guidance.

A total of 18 Shrike variants have been identified, this proliferation being largely due to the relatively narrow bandwidth of the early-model seekers: each new threat seemed to demand a revised seeker and a new model of the missile. By the time the final AGM-45A-9 and -45A-10 models were built, their seekers covered a greater frequency range than all the previous versions combined; the anti-radar missile had finally come of age.

The Wild Weasels were not the only Shrike platforms: the round also served on USN types such as the A-4 Skyhawk, A-6 Intruder, and A-7 Corsair II. Nor did its career end with the arrival of newer missiles. It is still in service on the F-4G and EF-111A, and has seen limited export service. Israel was supplied with several hundred configured for anti-"Fan Song" (SA-2) duties, and used these during the 1973 Middle East War, while some still serve on IDF-AF Phantoms and Kfirs. One Shrike round even created a diplomatic incident during the Falklands War by arriving in Brazil under the wing of an RAF Vulcan which had been forced to seek an emergency landing following a flight-refuelling problem. During the same conflict several were fired in anger during attacks on an Argentinian early-warning radar, but no direct hit was scored.

F-105G Wild Weasels

Even before the F-100Fs had reached southeast Asia, work was under way to convert 86 F-105F two-seat Thunderchiefs into Wild Weasels. The Republic fighter had the range and payload for deep strikes into hostile territory, and was fitted with avionics similar to those of the F-100F, plus underwing ALQ-71 or -87 jammers for self-protection. The prototype F-105 Wild Weasel flew in early 1966 and more than 20 were stationed in southeast Asia by the end of the year. Longer-ranged and faster than the F-100F, they were able to blunt the effectiveness of North Vietnamese SAM operations.

The aircraft clearly had the potential to carry a much more refined avionics fit, so the aircraft were later retrofitted to the definitive F-105G standard, receiving a range of avionics which made the -100F seem crude by comparison. By dividing the internal systems of the ALQ-101 between two fairings mounted on either side of the fuselage, Westinghouse created the ALQ-105 self-protection jammer, while the ALR-46 became the primary anti-radar sensor, locating targets operating between 2 and 18Ghz. The first rebuilt aircraft were in front-line service by the end of 1969. The F-105 Weasels started life armed with Shrike, but even before the definitive F-105F was deployed, a new anti-radar missile had appeared to supplement the -45.

AGM-78 Standard ARM

The General Dynamics Pomona AGM-78 Standard ARM was based on the naval RIM-66A Standard SAM, and has a maximum range of greater than 15½ miles (25km). A classic example of a US crash programme, it was rushed into service to supplement the shorter-ranged Shrike. Studies began in 1966, first flight was in 1967, and production of the -78B service version was under way by 1968. The -78C, -78D and -78D-2 followed in 1969, 1971 and 1971 respectively.

One little known variant was the ship-launched RGM-66D. Developed as an interim anti-ship missile for the US Navy, this was fitted to several US Navy destroyers, frigates and gunboats, plus three Imperial Iranian Navy destroyers. Until the arrival of the AGM-84 Harpoon anti-ship missile the RGM-66 provided a limited over-the-horizon missile capability, homing in on its victims by means of their radar transmissions; it could also be used in the SARH mode.

F-4G Wild Weasels

By the time the AGM-78D was fielded, a new Wild Weasel was available to carry it: the F-4G. Creation of a Weasel version of the Phantom was an obvious development, and work on the first conversions started in 1968. A total of 35 F-4Cs were

Above: AN F-4G Wild Weasel demonstrates the range of anti-radar stores which may be fitted. These are (from left) an AGM-45 Shrike, AGM-78 Standard, ALQ-119 jamming pod, AGM-65 Maverick, and AGM-88 HARM.

reworked as interim-standard anti-radar aircraft, receiving the ALR-46 radar homing and warning system and ALQ-119 dual-mode jammers for self-protection. These aircraft were delivered in 1969, but saw only limited service. Being based on the Phantom, the F-4G was more manoeuvrable than the F-105F, so it had a better chance in air-to-air combat if jumped by MiGs.

For the definitive F-4G, currently the USAF's front-line Weasel, a batch of 116 selected F-4E fighters with long fatigue lives were rebuilt at the service's Air Logistic Centre at Ogden AFB, Utah. Stripped of their 20mm cannon and fighter avionics, these aircraft were fitted with the advanced APR-38 radar homing and warning system, plus self-defence systems such as the ALQ-119 or -131

jamming pods and ALE-40 chaff dispenser. Deliveries began in April 1978.

Like all other US EW aircraft, the F-4G is being upgraded to keep abreast of new threats. A $200 million modification programme is extending the frequency coverage of the APR-38, the work being handled by E-Systems.

Iron Hand

Another Standard-armed anti-radar aircraft was fielded in 1968, but this was not a Wild Weasel. The threat posed by North Vietnamese SAM systems resulted in 19 A-6As being modified to the AGM-78 Standard-armed A-6B anti-radar version. Three different configurations of these Iron Hand SAM-suppression aircraft were deployed in small batches to supplement the strength of regular A-6 units. These aircraft retained varying degrees of strike capability, and at the end of the Vietnam War the surviving examples were modified to the A-6E standard.

AGM-88 HARM

Development of the Texas Instruments AGM-88 HARM (High-speed Anti-Radiation Missile) has taken more than a decade. The project was authorized in 1972 by the US Navy, but the USAF joined the programme three years later. Following

Below: General Dynamics has offered the F-16B Fighting Falcon in the HARM- and Shrike-armed Wild Weasel version shown here.

Above: The fuselage and wing of a warheadless HARM carved their way through the reflector and feed of this target antenna during a 1980 trial at China Lake.

Above left: Final moments of a HARM attack against a ship-mounted test antenna.

studies, Texas Instruments was given a development contract in May 1974, and 25 advanced development missiles were subsequently tested.

Trials apparently went well, but it was decided in January 1977 that the programme should remain at the development stage while the design was reworked to give greater frequency coverage and manoeuvrability. Full-scale development started in early 1978, with low-rate production of 80 missiles starting in FY81. Full-scale production of HARM was finally agreed by the DoD early in 1983, and a total production run of more than 17,000 rounds for the US services is planned. The USN will buy around 8,000, while the USAF has scaled its planned procurement down from more than 14,000 to only 9,000.

Initial deployment stàrted in 1985 on US Navy A-7E Corsair II and USAF F-4G Wild Weasel aircraft, but the type will also be fitted to USN F/A-18 Hornets and A-6 Intruders. Despite the competiton from rival British and French weapons, HARM could also be chosen as a NATO-standard weapon. Given a large enough order from NATO allies not

committed to other anti-radar missiles, HARM could be licence-built in Western Europe.

Production HARM rounds are reported to cost more than $390,000, a figure which has prompted urgent moves to devise a cheaper alternative. Studies of new versions of existing anti-radiation missiles have been carried out, but the Services elected to stick with HARM.

The most obvious area for cost-saving was the highly-sophisticated seeker, which accounts for around 50 per cent of the missile's cost. A broadband unit of advanced design, this can cover all threat frequencies, removing the need for customized missile versions. Whatever the threat, AGM-88 can cope. A new AGM-88B version of HARM with a less expensive seeker is now scheduled to replace the current missile on the production line in 1989, a move which is expected to bring the cost of HARM rounds down by some 20 per cent.

AGM-65D Maverick

Another new weapon in the F-4G's armoury is the AGM-65D IIR (imaging infra-red) version of the Maverick air-to-surface missile, which entered low-rate production in April 1983. More than 26,000 of the earlier TV-guided AGM-65A and AGM-65B versions were built. Although the high accuracy was ideal for engaging radar vans and SAM launchers, these were daylight-only weapons which relied on good visibility. The new -65D can be used at night or in conditions of poor visibility, and is not confused by smoke.

A total of more than 60,000 rounds are likely to be purchased, despite the -65D's hefty price tag of around $75,000. A number of modifications have already been carried out to cut the cost, thanks to the introduction by Hughes of new lower-cost technology in the missile's guidance and control sections.

Israeli ARMs

Israel used indigenously-developed anti-radar missiles to knock out Syrian SAM sites during the opening stages of the 1982 invasion of Lebanon. RPVs were sent in over the Beka'a Valley to act as decoys, and once these had induced Syrian SA-6 and SA-8 missile crews to begin transmitting, strikes were carried out with anti-radar missiles.

According to Israeli sources, the weapon is a bombardment rocket in the class of the Vought MLRS, although US Intelligence sources claim that air-launched missiles are used. The missile may be new, but the homing head is apparently that used by the AGM-45 Shrike. It is not clear whether these have been removed from Shrikes delivered by the US Government, or whether an alternative source has been found. Seekers might have been ordered from the US manufacturer or built in Israel with or without a manufacturing licence.

Air-to-air ARMs

Anti-radar missiles have been considered for the air-to-air role, but none ever saw service. The appropriately named XAIM-97A Seekbat was intended for use against the MiG-25 "Foxbat". Combining the airframe of the standard missile with a higher-performance rocket motor able to boost the round to "Foxbat"'s operating altitude and a specialized seeker designed to home in on the I-band radiation from the type's 600kW "Fox Fire" radar, Seekbat was part of the US response to what was sometimes depicted as a Soviet "Wonderplane".

The smaller Hughes Brazo started life as a US Navy weapon intended to attack hostile fighters by homing on their radar emissions. Flight tests were carried out using Sparrow airframes in 1974-75 before this project too was abandoned.

Sidearm

The most recent US anti-radar missile programme is Sidearm. Under this project the US Navy plans to deploy AIM-9C Sidewinder rounds rebuilt as short-range anti-radar missiles. Withdrawn from service due to unreliability, the -9C was the only version of Sidewinder to use radar guidance. Approximately half the 2,000 rounds built by Motorola are still in storage, and available for rework. About half the components

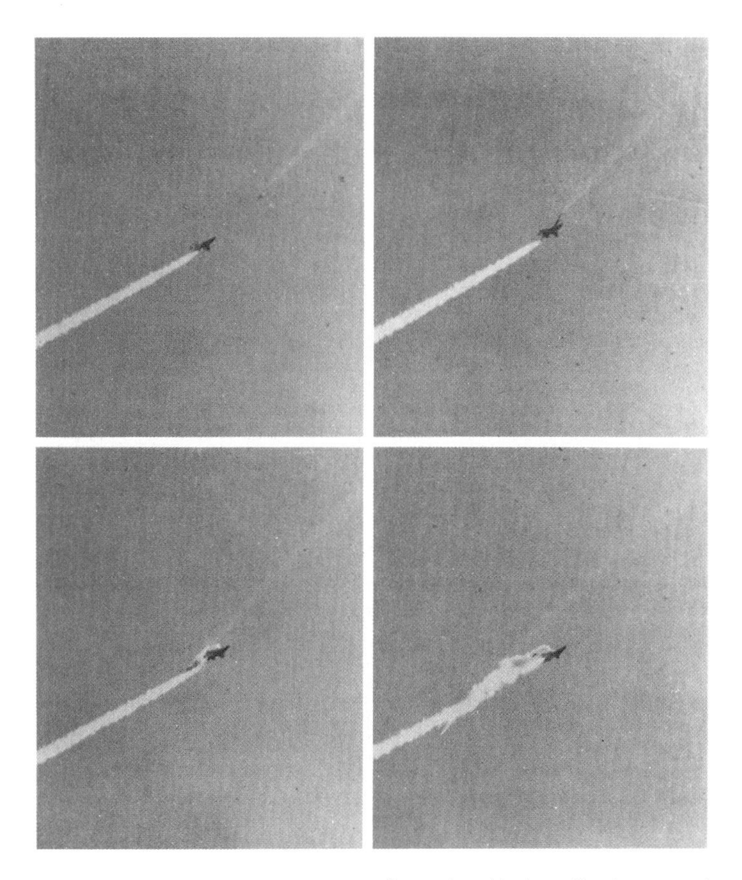

Above: The US services have never deployed an air-to-air anti-radiation missile, but this 1975 test of the Hughes Brazo showed that the concept could work. This was a long-range head-on shot from an F-4D.

in the seeker would be removed and replaced to create a passive homing missile suitable for use against the radar acquisition and guidance systems of new Soviet surface-to-air missile systems (probably the SA-13).

Following successful flight tests of modified rounds, the USN regards the concept as proven, and requested funding for the first production batch in Fiscal Year 1985. Motorola is expected to bid for the task of rebuilding these elderly missiles for their new role, but other companies known to be interested include Ford Aerospace, General Dynamics, Hughes, Raytheon, and Texas Instruments. Sidearm will be carried by smaller attack aircraft such as the AV-8, A-4 and AH-1.

Stand-off jammers

In addition to aircraft designed to directly attack hostile radar, the US Air Force and Navy also operate dedicated stand-off jamming aircraft able to escort attack formations to their targets, or to stand back well out of range of the defences and pour their massive jamming power towards threats in hostile airspace.

A US Marine Corps requirement for dedicated EW aircraft of this type was met in the short term by the EA-6A version of the Grumman Intruder. Introduced in 1963, this served as an interim design, the 27 built comprising a mixture of 6 modified A-6As and 21 new-build aircraft. The EA-6A retains a partial strike capability, but its main task is to counter hostile radars. Avionics include the ALQ-31 and -76 noise jammers, the ALQ-53 track breaker, and the ALQ-86 ESM system used to detect targets and to gather elint data. Deliveries were completed in 1969, and the type was in front-line service until 1978.

EA-6B Prowler

Production deliveries of the more drastically modified EA-6B Prowler started in 1971 and more than 100 have now been delivered. By stretching the basic airframe by 54in (137cm) Grumman was able to accommodate an extra two seats for specialist EW operators, and to pack within the fuselage and on underwing pylons the complex avionic units of the ALQ-99 jamming suite.

Designed for stand-off or escort jamming missions, this complex installation was created by an industrial team headed by AIL, and including AEL and IBM. Five external pods each contain two transmitters, a receiver and the associated antennas for these, while other equipment is located in the fuselage and in a fintop fairing. The system can function in automatic mode with the operators acting as monitors, or in manual mode.

With such extensively modified structure and avionics, the EA-6B costs around $40 million, a sum that would buy two A-6E Intruders, but if the latter ever have to go into combat, each Prowler will only have to save one or two A-6Es to make the programme seem very worthwile.

Given the rapidly evolving nature of the EW threat, equipment must be regularly upgraded in order to maintain its effectiveness. Even the most complex and flexible systems such as the ALQ-199 require regular upgrading. USN EA-6B Prowlers already carry an ICAP (improved capability) version, and they are

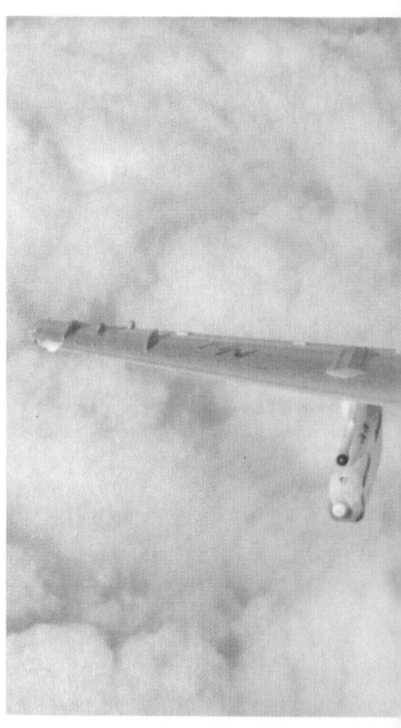

Above right: US Navy EA-6B Prowler with ALQ-99 pods on the outer wing pylons and external fuel tanks inboard.

Right: Rear view of an EF-111A showing the fintop ALQ-99 radome, plus smaller EW fittings on either side of the jetpipes.

Above: The Tu-16 "Badger-H" chaff-laying aircraft carries internally mounted dispensers.

scheduled to receive the EXCAP (expanded capability) version, which will cover a greater range of frequencies and have new deception modes.

EF-111A Raven

The increasing age of the Douglas EB-66 electronic warfare aircraft used in Vietnam led the USAF to rework obsolescent F-111A fighters to take over the stand-off and escort jamming role, and Grumman was awarded a contract in 1975 to convert a total of 42 to the EF-111A Tactical Jamming System (TJS) configuration. The aircraft is structurally unchanged and retains its existing engines and troublesome inlet system, but the lower fuselage and weapons bay are packed with jamming equipment, including an ALQ-137 ECM system for self-protection. The cockpit is also reworked for the new role, the right-hand crew member becoming an EW operator.

The USAF did not develop its own offensive jamming suite, but adapted the ALQ-99 from the Navy's EA-6B Prowler. (Purchase of the latter had been investigated as an alternative to the EF-111A programme, but the USAF concluded that the aircraft lacked the speed and endurance needed for use in Western Europe.) The Prowler avionics were repackaged to fit the F-111A's fuselage and ventral fairing, and automated so that they could be handled by the two-man crew. Frequency coverage of the resulting ALQ-99E system is reported to be less than that of the Naval system, which covers an additional higher-frequency band.

Operational testing in 1978 highlighted a number of deficiencies in the modified system, but these were soon cleared up, and approval for full-scale production was given in 1979. The first EF-111A Raven was delivered to the USAF in 1981, and aircraft were deployed to Western Europe in 1984. The final example is scheduled for delivery in late 1985.

The ALQ-99E will serve only as an interim jamming equipment. From 1988 onwards it will be replaced by a new Eaton/AIL system designed to be as easily retrofitted by replacing existing black boxes with new units of similar shape and function.

Soviet systems

There is no known Soviet equivalent to the Wild Weasel aircraft, and for a long time the Soviet Union apparently lacked Shrike-style anti-radar missiles. The only weapons thought to have the ability to home on radar transmissions were passive homing anti-ship versions of elderly winged

EF-111A avionics suite

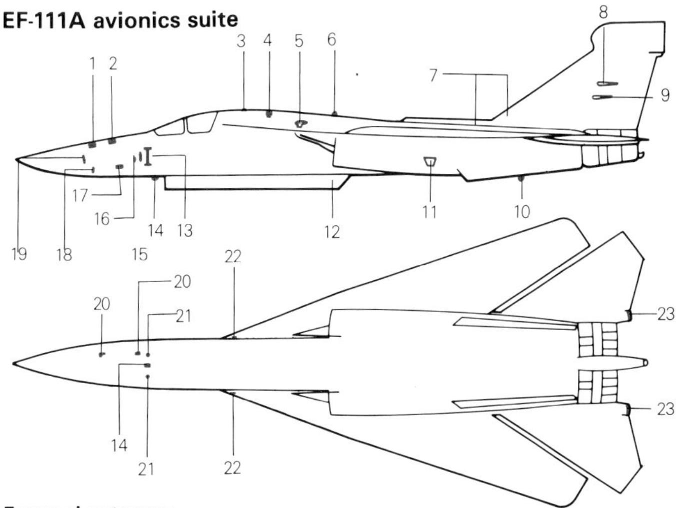

External antennas

1 Glide slope. **2** ADF. **3** IFF (upper) and UHF data link. **4** Radio beacon set. **5** ALQ-137 low/mid/high receiver and ALR-62 forward transmitter. **6** UHF No 1 and Tacan upper. **7** HF. **8** ALQ-99 band 1 (2). **9** ALQ-99 band 2 (2). **10** IFF lower. **11** ALQ-99 bands 1 and 2 (2). **12** ALQ-99 bands 4, 5/6, 7, 8 and 9. **13** Localizer (2). **14** UHF No 2 and Tacan lower.
15 ALQ-137 low-band transmitter. **16** ALQ-137 mid-band transmitter. **17** ALQ-137 omni mid-band transmitter. **18** TFR (2). **19** Navigation radar. **20** Radar altimeter. **21** ALQ-137 omni low and mid-band transmitter. **22** ALQ-137 high-band transmitter. **23** ALQ-137 high-band receiver and transmitter.

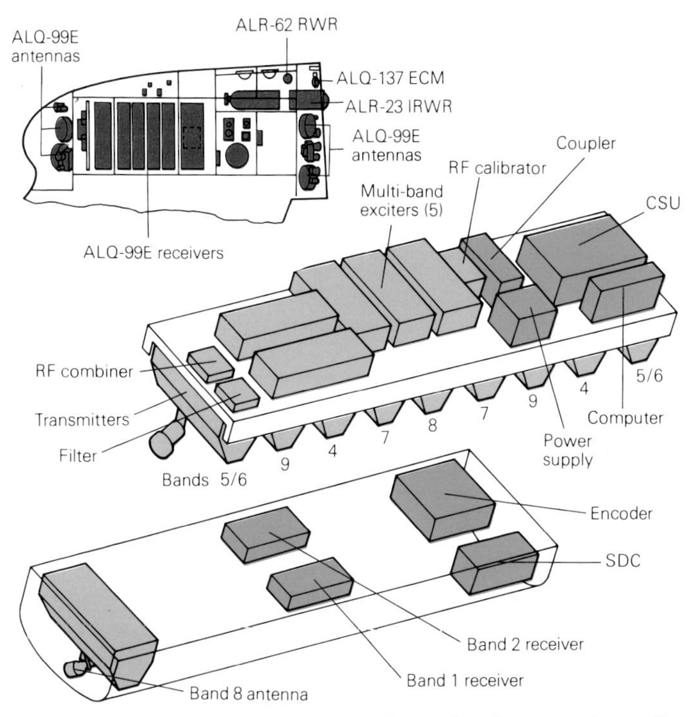

Fin-tip pod (top), ventral canoe radome (centre) and weapon bay pallet

EF-111A stand-off jamming role

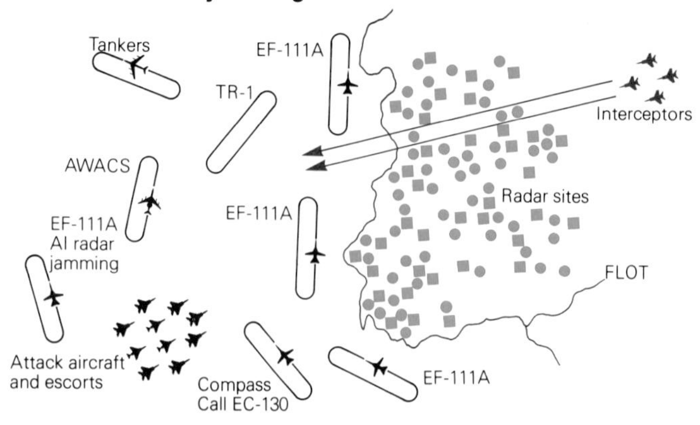

Above: Orbiting at high altitude, the EF-111A could jam the radars of enemy air defence systems and interceptors. The threat posed by such tactics may account for recent reports that the long-range SA-5 has been stationed in East Germany.

missiles such as the AS-1 "Kennel" and AS-5 "Kelt". The first modern Soviet ARM is apparently the AS-9 carried by the Su-24 "Fencer", but this is a relatively large missile with a maximum range of 50-56 miles (80-90km). Use of the US DoD designation AS-15 for the new Soviet air-launched cruise missile suggests the existence of hitherto undisclosed AS-12, -13 and -14 weapons, some of which might be smaller ARMs in the AGM-88 class.

Several dedicated jamming aircraft are in Soviet service. The Yak-28 "Brewer-E" entered service in 1970. Jamming equipment is carried internally, radiating via a large belly-mounted antenna, and a series of antennas in the nose section. Pods for chaff rockets are carried beneath the outer wings.

The Tu-16 "Badger-H" dedicated EW aircraft has a powerful installation of internally mounted jammers, plus ventral chutes and ejectors for dispensing chaff, flares and even expendable jammers. These ventral fittings are replaced on the "Badger-J" EW aircraft by additional antenna fairings.

Operational since the early 1970s, the Antonov AN-12 "Cub-C" was one of the first dedicated jamming aircraft to carry power-managed jammers. Palletized jamming equipment reported to cover "at least five wavebands" is located beneath the floor of the main cabin, while additional electrical generators mounted within the cabin provide the electrical power. Pods faired into the forward fuselage and underside of the aircraft presumably house antennas, as does the tailcone radome which replaces the tail gunner's position standard on AN-12 transports.

Anti-radar Tornado

No other nation currently operates specialised anti-radar strike aircraft, although the West German Luftwaffe is very keen to do so. Under a contract from the Federal German Ministry of Defence, MBB is working on methods of adapting Tornado to new tactical conditions and weapons, updates which will keep the aircraft combat-effective in the face of the likely Warsaw Pact air defences of the late 1980s and 1990s.

Modifications due for introduction on Block 5 aircraft and retrofitted to earlier models will allow German Navy Tornados to carry and launch the AGM-88 HARM. Aircraft will be fitted with a 128K central computer having twice the power of the current unit, and allowing the use of a MIL STAN 1553B databus. The new US DoD-approved ADA computer language will be used operationally for the first time in a more advanced pattern of missile-control unit, while

EF-111A close-in jamming role

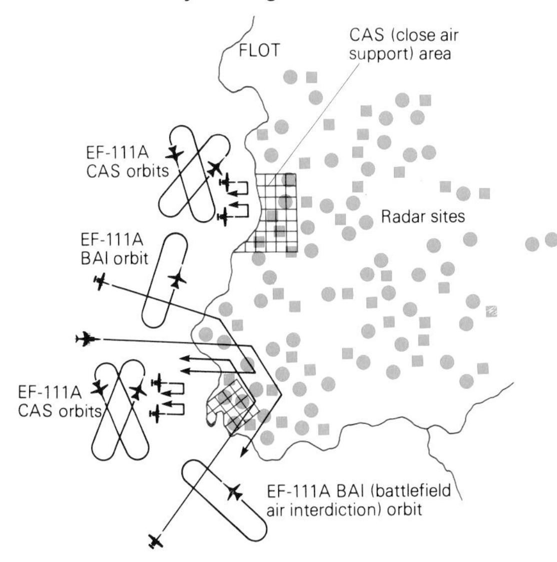

Above: Flying close to the FLOT (Forward Line of Troops), EF-111s could neutralize SAM and AAA acquisition and tracking radars, but would face the threat of long-range SAM or fighter attacks. Relying on high speed, a MiG-25 "Foxbat" might be able to get close enough to launch "home-on-jam" missiles.

EF-111A penetration escort role

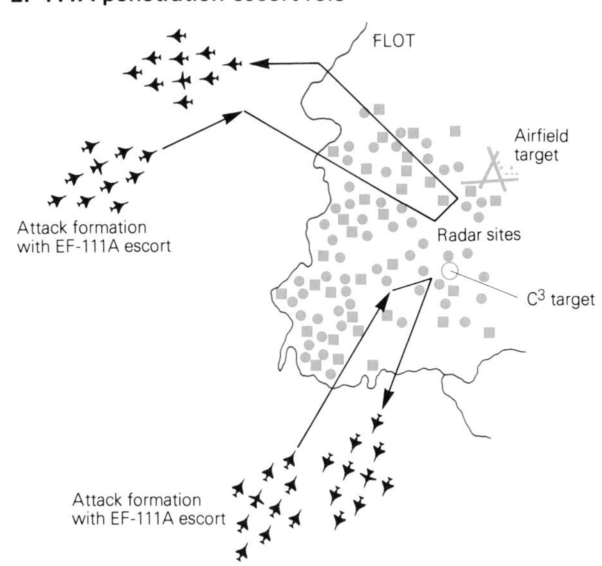

Above: In the escort role, a single EF-111 flying within a strike formation could protect the entire force. This exposes the aircraft to ground fire and medium and short-range SAMs, so could prove more dangerous than stand-off jamming.

electronic warfare systems such as the ECM transmitter, chaff and decoy pod and radar-warning receiver will be integrated into the overall avionics, allowing navigation and EW data to be combined and integrated into tactical information. Radar air-to-air and air-to-surface operating modes will be extended and refined to improve resistance to ECM, navigation, and weapon delivery.

This equipment standard forms the basis of the Tornado ECR (Electronic Combat and Reconnaissance) proposed to the Luftwaffe. This would carry a new Emitter Location System in place of one of the 27mm Mauser cannon, plus pod-mounted reconnaissance equipment and pod-mounted escort/stand-off jammers. The latter will probably be based on parts of the ALQ-99. The service has a firm requirement for the aircraft, but the decision on how many to buy and when had not been taken by mid-1985.

Alarm

At one time, the UK seemed likely to choose HARM as its long-awaited anti-radar weapon. The RAF has long needed a missile of this type, and repeated procurement delays plus the need to get weapons into service as soon as possible seemed to favour an order for the US missile. However, the UK MoD surprisingly decided to develop the indigenous British Aerospace Dynamics ALARM (Air-Launched Anti-Radiation Missile). Now under development for installation on Tornado IDS interdictors, it could also be fitted to the RAF's Jaguar, Harrier and Harrier II aircraft.

Being smaller than HARM, and only slightly larger than Sparrow, Alarm will be fitted on the sides of the aircraft's weapon pylons, supplementing rather than replacing existing stores, so it does not greatly reduce an aircraft's normal weapons load. Being completely autonomous, it requires little modification to the launch aircraft.

The seeker is programmed with two sets of data – a library of the emitters likely to be encountered, and the threat priority assigned to each. This may be done on the ground, or even in flight should the aircraft be diverted to a new mission or target. Given this data, the seeker does not need to be cued by the aircraft's RWR, but can detect and identify its own targets.

After release at low level, Alarm can operate in one of three modes. Direct attack involves the round flying directly toward the target, the flight profile adopted by weapons such as Harm. For attacks against area targets, indirect mode may be

Above: BAe's Alarm has a relatively small warhead, so it must rely on high accuracy in order to score a kill.

selected, in which case the missile climbs to around 40,000ft (12,000m), deploys a parachute, and begins to search for a suitable target. Once a target is selected, the parachute is released, and the missile starts an unpowered dive on to its chosen victim.

A recently added dual mode allows rounds launched in direct mode to switch to indirect if they fail to locate the original target. The indirect and dual-mode profiles do not require the seeker to be locked on before launch. If the target has not been acquired after a pre-defined flight time, missiles operating in dual mode will opt to attempt an indirect-mode attack.

Flight tests of Alarm are due to begin in early 1986, leading to a service entry date of 1987. BAeD has taken the risk of a fixed-price contract worth almost £300 million, and plans a production run of 2,000. This figure assumes large-scale export success, despite the fact that previous British air-launched missiles have attracted few customers.

ARMAT

Europe's only other anti-radar missile programme is the Aérospatiale ARMAT (Anti-Radiation MArTel). This combines the proven airframe of the original AJ37 anti-radar version of the Martel air-to-surface missile with a new Electronique Serge Dassault (ESD) homing head. Back-up INS

Above: The anti-radar version of Martel (seen here on an RAF Buccaneer) does not seem to have been very successful. In the Falklands War, the RAF relied on US-supplied Shrikes.

guidance may be fitted to deal with radars which attempt to confuse the missile by switching off their transmitters.

Wild Weasel countermeasures

Some countermeasures to Wild Weasel operations are possible. Back in the mid-1960s, North Vietnamese missilemen learned ways of coping with the Weasels and their Shrike armament. By coordinating their operations with those of early-warning and general surveillance radars, crews were able to use data from the latter to track the position of US strike formations, eliminating the need to turn on their own sets for target-tracking purposes. With their SAM-guidance radars running, but transmitting into a dummy load instead of the antenna, crews could wait until the last possible moment to start radiating, acquire the target, and launch a missile.

Radars could avoid attack by anti-radar missiles such as Shrike or Standard by the simple expedient of switching off their transmitters, but the weaponry of the F-4G should blunt this tactic in any future conflict. The high speed of the HARM missile gives SAM crews less time to react to a Wild Weasel missile launch, while the imaging infra-red seeker of the AGM-65D Maverick allows the Weasel to continue its attack by launching a round able to home on the residual heat emitted by now-silent radar.

PLSS

Another possible counter to anti-Weasel shutdowns is the Precision Locator Strike System (PLSS) intended for service on the Lockheed TR-1 reconnaissance aircraft. Delayed by Congressional funding cutbacks, PLSS now lags behind the TR-1 in development, but a production go-ahead in 1985 seems likely.

Based on technology tested in 1972 and 1975 but never deployed operationally, PLSS involves maintaining three aircraft orbiting on station in friendly airspace. These observe radar signals originating on the other side of the front line, noting the precise time of arrival and direction of each signal. The position of each aircraft at that instant can be determined by using ground-based DME (Distance Measuring Equipment), so the exact location of the emitter may be deduced.

Above: This test rig simulates the wing of the Lockheed TR-1, but uses metal mesh rather than a metal skin in order to reduce weight and wind resistance. It is being used to measure the reception pattern of podded Distance-Measuring Equipment antennas for the Precision Locator Strike System (PLSS).

Like the E-3 Sentry (AWACS), PLSS-equipped TR-1s may be used to vector friendly aircraft, in this case tactical fighters tasked with destroying the threat radars. If the fighters carry DME-guided weapons such as the DME version of the GBU-15 glide bomb, the latter could be released at stand-off range then guided to impract by the PLSS system. An accuracy in the order of 75ft (25m) or better seems probable.

Future systems

Anti-radar missiles currently under development are likely to serve through the 1990s and into the early years of the next century, but they could be supplemented by newer missiles designed to take advantage of some ideas still in the laboratory.

Radars working at some of the lower frequencies favoured by Soviet designers are difficult to home in on: they can be detected easily, but the angular resolution needed to score a hit requires an antenna larger in diameter than a missile may conveniently carry, since an antenna's beamwidth decreases with decreasing frequency.

Some form of synthetic-aperture seeker may prove effective in attacking such difficult targets. Design studies have already investigated ways of firing a two-round salvo and using the spacing between the seekers on the individual rounds as the baseline for synthesizing a higher-resolution antenna. This involves maintaining a data link between the two seekers, and studies have even considered linking the two

Sanctuary

Below: The Sanctuary system would mount radar transmitters in TR-1s or even spacecraft.

Front-line radar sites would be passive and invulnerable to anti-radiation weapons attacks.

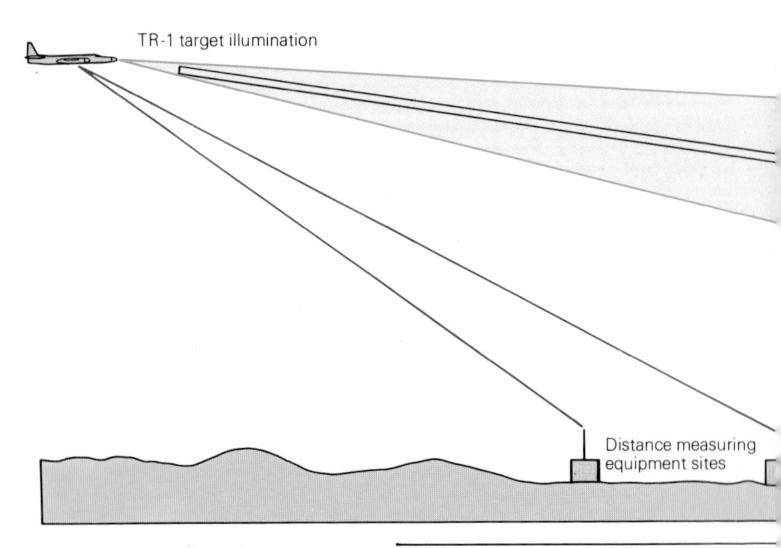

missiles together by an optical fibre throughout their flight.

The current F-4G Wild Weasel aircraft are rebuilds, with a limited service life, and while there has been little public discussion of Wild Weasel F-15s, the type seems the only logical replacement for the existing aircraft. By 1988 the USAF plans to take delivery of its first two-seat F-15E Eagles, multirole fighters based on the earlier experimental Strike Eagle night/all-weather ground attack fighter.

The new aircraft will have a Hughes APG-70 radar with synthetic-aperture air-to-ground modes, the LANTIRN multisensor pod, and an advanced flight-control system, and will be able to carry air-to-ground munitions such as iron bombs, retarded bombs, smart bombs, cluster munitions, nuclear weapons, and air-to-ground missiles such as Harm and Maverick. The basis of a new Wild Weasel is clearly there, so the F-4G seems assured of a long-term replacement. Teamed with PLSS-equipped TR-1s, a "Wild Eagle" fighter would be a formidable anti-radar weapon.

PLSS operation

Below: PLSS-equipped TR-1s can locate hostile emitters, then guide strike aircraft or missiles.

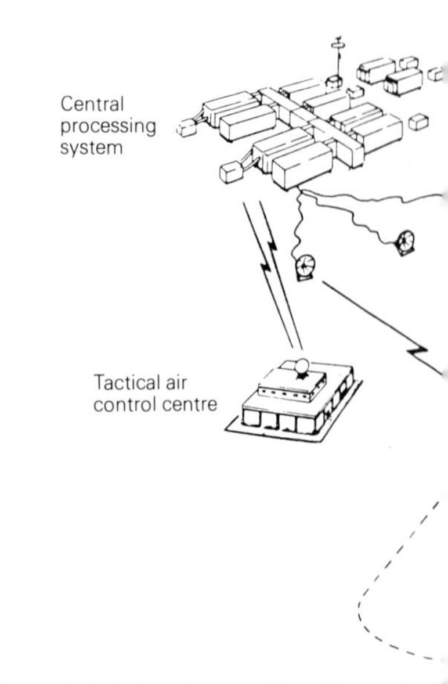

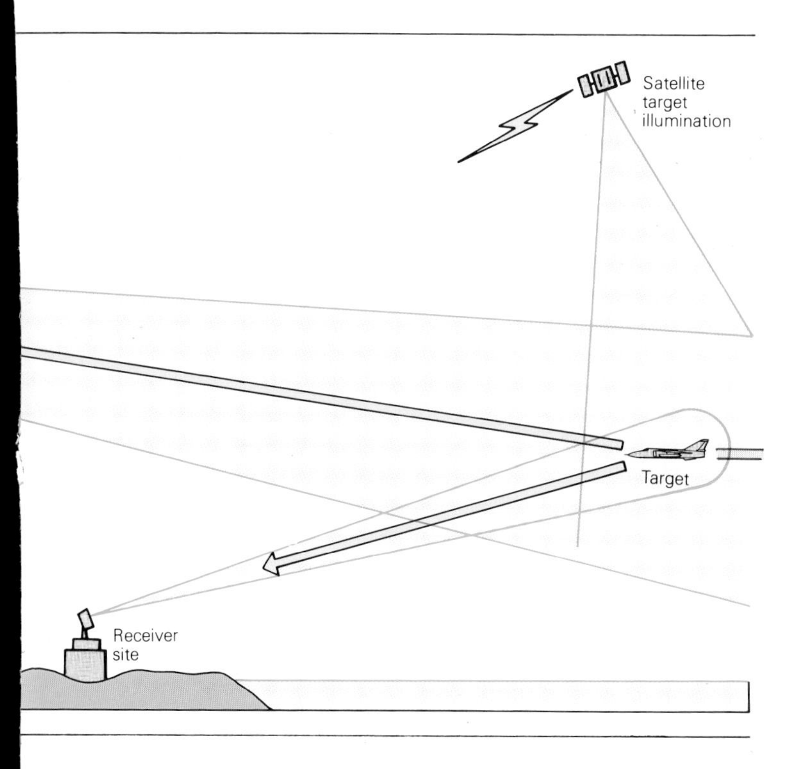
Satellite
target
illumination
Target
Receiver
site

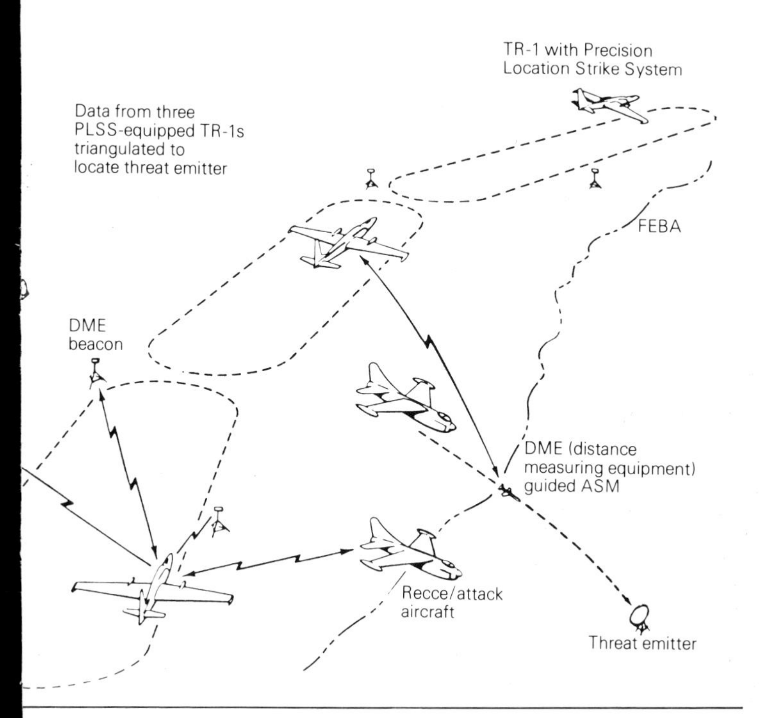
TR-1 with Precision
Location Strike System
Data from three
PLSS-equipped TR-1s
triangulated to
locate threat emitter
FEBA
DME
beacon
DME (distance
measuring equipment)
guided ASM
Recce/attack
aircraft
Threat emitter

Electromagnetic Pulses

GROWING reliance on electronics and computers has produced new areas of vulnerability in modern defence equipment. The additional complexity is certainly resulting in smarter and more effective systems, but has also created a new and unconventional threat – the prospect of electronic systems being damaged irreparably by powerful electromagnetic pulses generated by nuclear explosions.

No type of system would be spared the effects of such damage. Unless specifically designed to cope with high-intensity pulses, everything exposed to an electronically lethal pulse would fail, from the soldier's manpack radio to the most complex missile, radar and command and control systems.

In addition to the well-known heat, blast and radiation effects from nuclear explosions, there is also in the case of high-altitude explosions an intense pulse of radio energy caused by gamma photons from high-altitude nuclear explosions liberating large numbers of electrons from atoms of atmospheric oxygen or nitrogen. Reacting with the earth's magnetic field, the liberated electrons move in spiral paths due to the cyclotron effect, and in so doing emit a short electro-magnetic pulse (EMP) of radio energy. The frequency spectrum of this EMP extends from 10kHz to 100MHz – exactly the range used by the majority of communications systems.

This pulse is of very short duration, but of intense power. A high-altitude nuclear blast of about one megaton yield would generate an electromagnetic field of several tens of kilovolts per metre. Although the pulse would last only some 200 nanoseconds, the instantaneous power level would be of the order of

Right: Probably the world's most famous electro-magnetic pulse simulator is that operated at Kirtland AFB by the USAF. The massive test platform is made from wood, so has minimal effect on the electronic characteristics of the massive pulse of energy created by the antenna wires surrounding it.

500 billion megawatts, sufficient to wreak havoc with electronic systems, particularly those using solid-state electronics.

Electrical fields induced within conventional metal oxide/silicon (MOS) semiconductor devices can alter their characteristics, ultimately causing a malfunction. Integrated circuits could fail completely, or else exhibit erratic or intermittent behaviour, while the striking of electrical arcs within the affected unit could complete the destruction. The more complex the circuitry, the greater the number of sensitive elements which could be damaged, while the faster the circuitry, the more sensitive the individual elements will be.

The very factors which make digital equipment more powerful and useful make it ever more sensitive to EMP attack. Conversely, older and simpler equipment based on the thermionic-tube (valve) technology will be least vulnerable to EMP — a fact which leads some observers to question whether the widespread use of tubes by the Soviet forces is entirely due to aging and obsolescent equipment.

Radiation hardening

The threat of radiation damage must be faced by the designers of any advanced military system. The solution adopted by the West is not a return to or retention of relatively simple tube-based electronics, but the protection of solid-state electronics from EMP, a technique known as "radiation hardening". Procurement agencies are increasingly specifying that new equipment and even some existing systems should be radiation hardened.

One simple technique is to mount sensitive electronics within a metal box which is electrically "sealed", and which will thus shield the contents from EMP. This is a good technique for upgrading existing equipment, but requires close attention to detail. Any openings such as hatches or covers must make a good electrical fit all around their edges, while electrical cables entering or leaving the box are fitted with pulse limiters and suppressors to prevent EMP-induced electrical surges from entering via the cables.

Equipment designed since the implications of EMP were thoroughly understood can use modern components designed to have a good resistance to EMP, and may use resonance-frequency suppression and other techniques.

EMP-resistant microchips

By using suitable techniques, designers of electronic microchips can give their creations good resistance to EMP. This can be done either by employing specialized manufacturing processes, or by careful choice of semiconductor material.

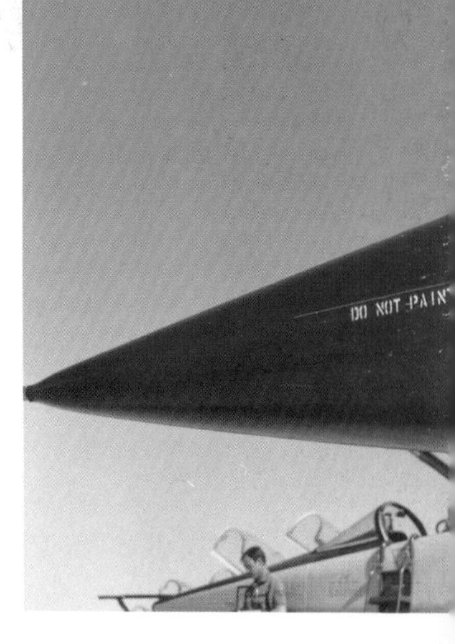

One example of this approach is seen in the Ferranti F100-L, a 16bit microprocessor designed for use in fast real-time systems. This component and its family of associated devices use a technique known as collector diffusion isolation (CDI) – bipolar transistor technology which is less sensitive to radiation-induced changes in the silicon material. Together with certain circuit-design techniques such as the use of feedback emitters, it confers good radiation hardness on the entire family of devices. Predicted hardness figures have been independently tested at the Atomic Weapons Research Establishment, Aldermaston.

One semiconductor material which promises significant improvements in radiation hardening is the use of gallium arsenide as a semiconductor material. Pilot pro-

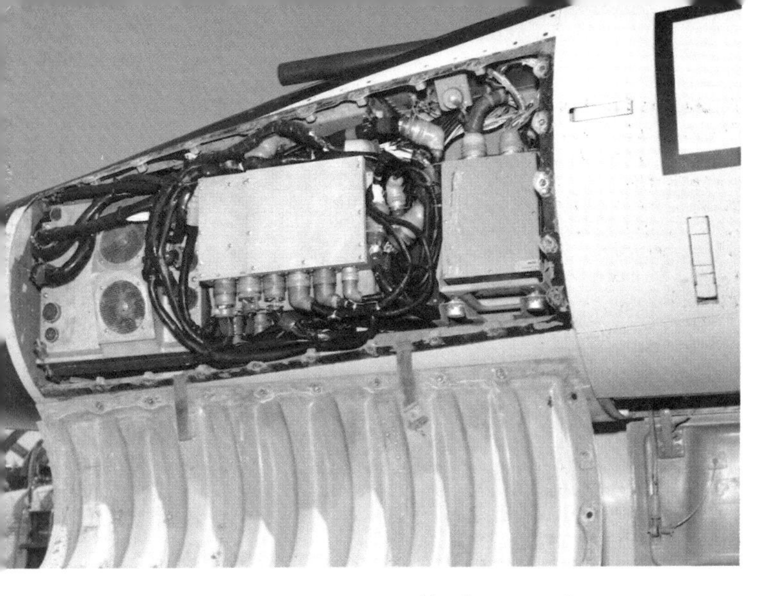

Above: If precautions are not taken, the electrical cabling in an aircraft (this is the Emerson APG-69 radar on test in an F-5E) could act as a receiving antenna for unwanted EMP energy.

duction of integrated circuits etched in gallium arsenide chips has already been demonstrated in the US. This has important potential applications in space-based systems.

Hardened command posts

In the mid to late 1980s, the US DoD will make a determined effort to improve the EMP survivability of the National Military Command System. The ground-based National Military Command Centre (NMCC) and the Alternate NMCC will be given greater information-processing equipment, and enhanced protection against EMP. Although useful during normal crisis-management, and of obvious service during the early stages of a nuclear attack, both are likely to be high-priority targets for the Soviet SS-18 force. The four Boeing E-4B National Emergency Airborne Command Post aircraft are all being hardened against EMP, as are the smaller EC-135 airborne command posts.

Left: Piaggio supplied these EMP-resistant equipment shelters to the West German Army for use in the 4-ATAF programme.

Hardness testing

The structure and avionic systems of the B-1B bomber are designed to absorb nuclear effects. Electronic units, cabling, and other sensitive components are all shielded against the effects of EMP by a combination of circuit hardening and electrical screening. Such hardened equipment must be tested in order to check its resistance to EMP. Since a nuclear explosion is hardly a suitable or convenient test instrument, simulation facilities must be used – equipments designed to flood a test area with a powerful EMP, or to inject suitable pulses into specific points of the equipment under test.

Facilities for carrying out such tests have been built in many locations such as Kirtland AFB (USA), AWRE Aldermaston (UK), and the Les Mureaux plant of Aérospatiale (France). These very large fixed simulators were built on the assumption that the equipment to be tested is capable of being moved to the test site. Some are designed to test small items such as electronic sub-units, while others can cope with complete missiles, AFVs or even aircraft – the largest equipment at Kirtland AFB has a wooden test platform able to carry an entire B-52. Mobile simulators have also been devised to permit the EMP testing of fixed installations such as command posts and communications centres.

Unconventional Threats

WITH the large-scale use of computers by the military, two rather odd classes of threat have emerged -- possible methods of extracting from a computer a copy of the classified information which is it handling. Neither is a threat to the computer systems embedded within modern aircraft, warships and radars, but each could be a hazard in research establishments and command and control centres.

Data leakage

First of these threats to be recognized is that of data leaking from computers by spurious radiation – a form of sigint which seems at first sight to border on the realm of science fiction.

The internal circuitry within computer equipment operates at frequencies within the normal range used for radio communications. As a result, the computer tends to radiate electrical noise as individual elements in its internal circuitry "broadcast" the data that they are handling. This stray radiation is at a very low level, but sufficient to cause concern – owners of portable business micro-computers are not permitted to use them aboard airliners, since the radiation could affect the plane's avionics. From the military standpoint, such unwanted emissions also provide a potential method of espionage.

Since all parts of a computer are radiating simultaneously and all are handling different data, the result is in theory a low-powered electronic cacophony. In practice, a skilled operator equipped with suitable receiving systems can pick out the vital data, and the computer user will have no way of knowing that he has been robbed. This is no theoretical intelligence-gathering technique. During one demonstration held in the US of such eavesdropping techniques, data being displayed on the terminals of a business computer was reproduced on another display in the building opposite.

The Tempest programme

To curb the military possibilities of this technique, the US National

Right: An EMP-resistant building under construction for the UK MoD. The walls are lined with metal, as will be the doors, with the completed building forming an electrically-sealed unit.

Security agency (NSA) has set up a programme known as Tempest. Equipment already certified under this scheme includes computers, printers, displays, and cryptographic equipment.

In order to be Tempest-rated, equipment must be designed to suppress unwanted radiation. This is partly a matter of arranging good electrical screening or shielding within the equipment, which may in turn be operated within a shielded room. One unclassified source has mentioned the possibility of electronic countermeasures such as the deliberate radiation of radio noise intended to swamp the signal which the eavesdropper is trying to detect.

Potential weak spots in an equipment's electronic screening include interconnecting cables and CRT display screens. Radiation from cables can be minimized by the use of optical fibres instead of electrical cabling. Transmission of data between items of computer equipment as pulses of light, rather than electrical pulses, allows information to be sent over long distances without the risk of it 'leaking'.

Display terminals are difficult to screen, since the CRT screen requires a cut-out in the metal enclosure surrounding the equipment so that the tube face will be visible. Moreover, the electrodes within the CRT which are used to direct the electron beam which forms the characters displayed on the screen are energized with high voltages modulated by the data which the user wants to keep secret. They are thus a good potential

source of compromising radiation, and must be screened with a glass/mesh window or some other method.

Secure rooms

In cases where highly classified data is being handled by computer, the entire system may be mounted in an electrically shielded room. A typical installation of this type is the computer suite of Marconi's electronic-warfare laboratory at Stanmore in England. Intended to handle much of the computer simulation work which takes place early in any ECM development programme to ensure that the hardware eventually created fully matches the operational requirement, this installation could expect to be a natural target for espionage activity.

One clue to the precautions being taken against hostile elint can be seen around the door and doorframe leading into the high-security computer room. Long seals made from tiny metal fingers designed to intermesh as the door is closed form the electronic equivalent of a draught excluder. When the doors are closed the computer room forms a signal-tight metal box — a fully-screened room within which the most sensitive information can be handled in complete security and safety.

Hacking

Electronic intelligence-gathering of data from computer systems is not restricted to the technology which Tempest rating attempts to defeat. The cinema film *War Games* released in 1983 highlighted another potential threat to military computer systems — the risk that unauthorized individuals or organizations may attempt to gain access to the machine and its data via the telephone. Many computers exchange data via normal telephone lines by using interface units known as modems (modulator-demodulators). In the US such electronic burglary — known as hacking — is the unofficial hobby of some home computer owners.

As hackers learn the telephone number of a computer, details are passed on to others via bulletin board systems (BBS) — home computers linked to the telephone network by modem, and run by computer enthusiasts as a method of exchanging information. Many boards will not accept messages and information related to hacking, but some keep such data in a "confidential" section which only trusted hackers will be allowed to consult. Others openly accept hacking information.

Computers connected to the telephone system normally rely on a system of passwords. When an outsider phones the appropriate telephone number, he will automatically be connected to the computer via its modem, but until he has identified himself by giving a password, the computer will refuse to accept his instructions.

Hackers in action

Once the number has been publicized, hackers will attempt to break through the computer's password system, leaving messages on BBS to report commands which the machine under attack has been found to accept. The messages which appear below document the opening stages of a hacking operation. The text has been slightly edited, to simplify the technical jargon and to protect the identity of the BBS, hackers and the computer being investigated.

It began by one hacker reporting an interesting phone number. "Try calling xxx-xxx-xxxxx. When you are connected, type LOG 123456, 7 — this gets you into what seems to be a college computer. I tried all variations but I have not yet cracked the password... Have fun. Please report results."

Other hackers were quick to join in. Further information was given the same day in two more messages. "The number gets you into the University of XXXXXXXX computer. PAD is obviously one facility. When you input CALL 40 it seems there must be a space between CALL and 40. Anyone had any interesting finds on the system?"

"The University of XXXXXXXX computer? It has several, and a lot are accessible over the telephone network. Keep going, you've a lot of machines to find yet. Where's the

Above: Stacked to the right of this Apple II computer are disk drives for data storage, and modems able to operate on US and European standards — all the equipment needed for hacking.

challenge in accessing academic machines anyway? They aren't aiming to achieve any great security."

Within a month, further instructions were being given. "Type HELP or HELP ADDRESS and as good as gold, it gives you a whole load of addresses to call. Just try CALL (SPACE) (ADDRESS)."

Once they have gained access to the computer, hackers can read confidential data. Most systems offer several levels of access, with any single password giving access only to the data which the legitimate user of the password required. Some hackers have the skill to install a "demon" — a small program which monitors the log-on procedure of other users, recording the passwords being used. Interrogating the demon at a later date, the hacker can obtain a list of all recently-used passwords.

Like the earlier craze for "phone freaking" (manipulation of the telephone system to obtain long-distance calls at local-call rate or even free), hacking is largely practised by young people and is seldom carried out with malevolent intent. Operating under colourful CB-style pseudonyms ("handles"), most hackers are content simply to read data files, or at worst to leave brief messages to inform legitimate users that system security has been breached. One well-known target for

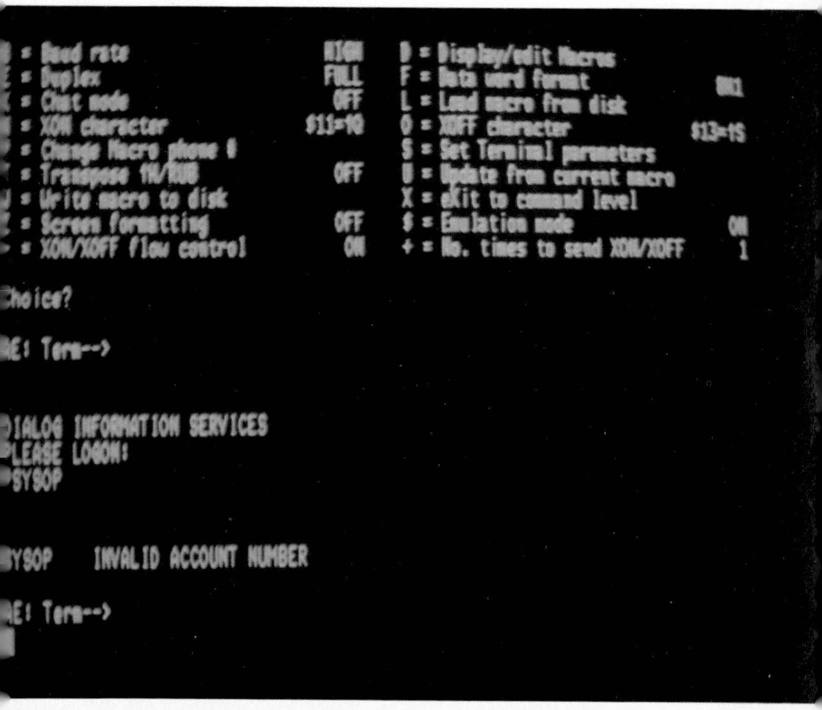

```
 = Baud rate                   HIGH   D = Display/edit Macros
 = Duplex                      FULL   F = Data word format            8N1
 = Chat mode                    OFF   L = Load macro from disk
 = XON character            $11=^Q   O = XOFF character           $13=^S
 = Change Macro phone #               S = Set Terminal parameters
 = Transpose ^H/RUB             OFF   U = Update from current macro
 = Write macro to disk                X = eXit to command level
 = Screen formatting            OFF   $ = Emulation mode               ON
 = XON/XOFF flow control         ON   + = No. times to send XON/XOFF    1

hoice?

E: Term-->


IALOG INFORMATION SERVICES
LEASE LOGON:
SYSOP


YSOP    INVALID ACCOUNT NUMBER

E: Term-->
```

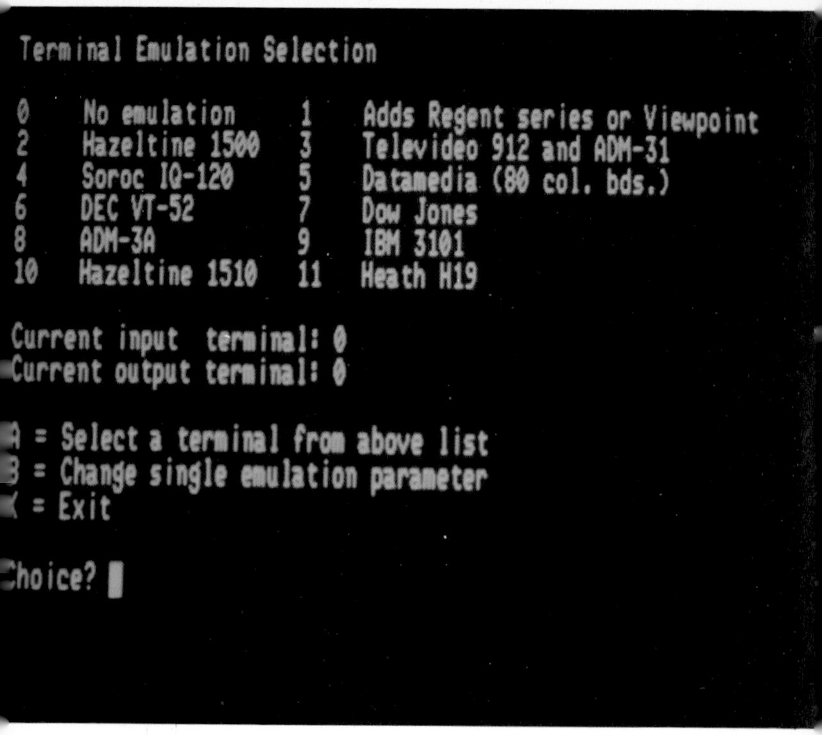

```
Terminal Emulation Selection

0    No emulation       1    Adds Regent series or Viewpoint
2    Hazeltine 1500     3    Televideo 912 and ADM-31
4    Soroc IQ-120       5    Datamedia (80 col. bds.)
6    DEC VT-52          7    Dow Jones
8    ADM-3A             9    IBM 3101
10   Hazeltine 1510     11   Heath H19

Current input  terminal: 0
Current output terminal: 0

A = Select a terminal from above list
B = Change single emulation parameter
X = Exit

Choice?
```

UK hackers is an electronic file maintained on the British electronic-mail system by the Duke of Edinburgh, consort of Queen Elizabeth.

Sinister applications

Phone freakers were more of a nuisance than a criminal threat, but hacking has more serious overtones. A small minority of hackers attempt to destroy data files, and some even try to disrupt the normal running of the computer they have accessed. The techniques used could be applied to various criminal ends – manipulating bank accounts or invoices, for example – and also for industrial or military espionage.

The degree to which military computer networks may have been penetrated by hackers is probably exaggerated, but some systems have been broken. The most common military "victim" is probably the Arpanet network which the US DoD uses to link many of its research contractors. One US hacker interviewed on British TV described how he tapped into a long file of seismic data originating from Norway. On checking the latitudes and longitudes mentioned in the data, he discovered that the corresponding positions lay deep in the Soviet Union. The information he had accessed was almost certainly data from a highly classified seismic system deployed to monitor Soviet underground nuclear tests.

Official reaction

So far, there have been no recorded instances of hacking skills being applied to espionage, or of any data obtained by hacking having been passed to an unfriendly power, but the possibility is clearly troubling the US authorities. In the summer of 1983 the FBI finally decided to get tough with the hackers, and 14 individuals had their homes raided and computers seized. Most were juveniles. Two months later, two California-based hackers were arrested and charged with having illegally accessed "very sensitive" computer data held by universities and agencies conducting DoD research.

Hacking has not been stamped out, but the incidents to date may actually have helped to improve computer security. Some organizations even employ hackers as consultants to advise them on computer security.

The US DoD certainly takes the threat seriously. Additional security devices are being added to existing telecommunications systems, while newly-issued design specifications for the latter now define suitable 'anti-hacker' measures.

The security measures applied to business computers are often rudimentary. Users are sometimes allowed to select their own password, a scheme which is fraught with danger since users tend to select something which they can easily remember. Computer security consultants have on record endless tales of wives' names, husbands' names, children's names, dogs' names, car registration numbers and similarly unimaginative passwords being guessed by unauthorized users.

In the film *War Games*, the teenage hero tries the name of the programmer's son as a password and promptly gains access to a Pentagon computer which controls the ICBM force, inadvertently triggering off a false attack warning. This might make good fiction, but military systems are unlikely to use such a simple password system.

To take a simple instance, if a user logs into a system by telephone, the system might not immediately accept the password, but break the connection, check in its records the telephone number of the user assigned that particular password, then remake the connection by dialling that number. This procedure ascertains whether the request for access came from the correct location and not from a hacker.

Above left: In the spring of 1984 a UK magazine openly published the procedure for connecting to Dialog, a Lockheed computer system. Here the system is seen rejecting a fake password.

Left: The hacker's equipment and software must be able to emulate many types of computer terminal. The intruder is thus able to electronically resemble a legitimate user.

Glossary

AGI Naval term for an intelligence-gathering vessel

AIM-US designation for air-to-air missiles

Algorithm Mathematical process for achieving a desired result

ALR-US designation for radar-warning receivers

Aluminised Coated with a microscopically thin layer of aluminium in order to create an electrically-conducting or optically-reflecting surface

Analogue Electronic system in which quantities are represented by electrical signals of variable characteristics

APQ-US designation for jamming systems

ASPJ Advanced Self-Protection Jammer

Backward-wave oscillator Electronically-tunable signal source based on the same principle as the TWT

Band Designation for a range of frequencies (for example, the I and J bands often used for radar stretch from 8 to 12 Ghz)

Bandwidth The range of frequencies over which an antenna, sensor or communications channel is designed to operate

Beamwidth Angle over which an antenna delivers most of its power, or receives most of its signal

Centroid The geometric mean of all the points contained in a shape

Chaff Strips of electrically-conducting material used to generate false radar echoes

Clutter Unwanted radar indications on a radar display; usually due to natural or man-made interference or echoes from the surface of the land or sea

Comint Intelligence gathering by monitoring communications

Conical scan The rotation of an antenna feed or other sensor around its nominal direction of "gaze" in order to measure aiming error

Continuous wave Radar signal which constantly illuminates a target, relying on modulation to determine range

CRT Cathode-ray tube (TV-style display screen)

CW Continuous wave (see above)·

Databus Electronic "highway" for the flow of data around a digital electronic system

DECM Deception methods of electronic countermeasures

Digital Electronic system in which quantities are as on/off signals coded to represent numbers

Dish Common term for a circular paraboloidal antenna

Dispenser Device designed to release chaff cartridges, flares, or other EW stores in a pre-arranged or remotely-controlled manner

Doppler effect The shift in a signal frequency due to the relative motion of the signal source and the observer

ECCM Electronic counter-countermeausres

ECM Electronic countermeasures

ELINT Intelligence gathering by the monitoring of radar and other non-communications signals

EMP Electro-magnetic pulse, normally a reference to the intense pulse created by high-altitude nuclear explosions.

EO Electro-optical

EOCM Electro-optical countermeasures

EW Electronic warfare

Expendable When used as a noun by EW specialists, this refers to chaff, flares or throw-away jammers

Ferret Platform equipped for sigint missions

Flare Source of intense heat (normally a pyrotechnic device) released in order to confuse IR seekers and sensors

FLIR Forward-looking infra-red

Frequency-agile Transmitter or other other emitter able to rapidly vary its output frequency from one scan of the antenna to the next

Frequency hopping A method of operating a radar or communications sytem by switching rapidly from one frequency to another hundreds or thousands of times each second according to a pre-arranged plan

CGHQ Government Communications Headquarters, the main UK sigint organization

GHz GigaHertz (Hertz x 1,000,000,000)

Hacker Strictly speaking this slang term indicates only a computer enthusiast, but is often used to indicate an individual who attempts to gain unauthorized access to a computer system

Hz Hertz (unit of frequency)

IIR Imaging infra-red

IR Infra-red

IRCM Infra-red countermeasures

JTIDS Joint Tactical Information Distribution System

kHz KiloHertz (Hertz x 1,000)

Klystron Source of microwave power based on a velocity-modulated electron beam

kW Kilowatt (Watts x 1,000)

LORO Lobe On Receive Only

Magnetron Commonly used high-power source of microwave energy

MHz megahertz (Hertz x 1,000,000)

Micron One millionth of a metre

Microwave Region of the spectrum between 1GHz and 300GHz

Modem Modulator/Demodulator — an item of electronics used to pass data down a communications channel

Monopulse Simultaneous measurement of target range and bearing using a single pulse

MTI Moving Target Indicator — a system which displays only moving targets, removing returns from terrain and other stationary features

MW Megawatt (Watts x 1,000,000)

NSA National Security Agency, the main US sigint organization

Paraboloid Shape formed by rotating a parabola around its own axis

Penaid Penetration aid — EW package carried on a ballistic missile to confuse defensive radars or missiles

Phase In electronic terminology, relative position in a regularly recurring cycle. Signals accurately matched in time and frequency are described as being "in phase"

Pulse-Doppler A sophisticated form of radar which can detect low flying targets from within a background of clutter

Pyrophoric Able to ignite spontaneously when in contact with air

Raster scan Method of building up a TV-style image on a CRT by scanning the image in a series of lines

Reticle An aiming mark or series of marks in the image plane of an optical system

Rint The gathering of intelligence by monitoring stray radiation

RWR Radar-warning receiver

PPI Plan-position indicator

PRF Pulse repetition frequency

Program Instruction for a computer

Sidelobe Unwanted secondary beam of an antenna.

Sigint General term for all form of intelligence gathering which involve the interception and analysis of electromagnetic radiation

Software One or more programs for a computer or other digitally-controlled item of electronics

Synthetic-aperture Radar technique by which a small antenna on a moving vehicle may simulate a larger unit in terms of resolution

Tempest Codename for a Western programme to reduce the amount of stray energy emitted by electronic systems

Track-while-scan Radar able to track one or more targets while continuing its search pattern

TWS Track-while-scan (see above)

TWT Travelling-wave tube (power source used in many modern radars)

Wide-band Device designed to operate over a wide range of frequencies

Window Area of the electromagnetic spectrum not attenuated by the gases and water vapour in the earth's atmosphere

OTHER SUPER-VALUE MILITARY GUIDES IN THIS SERIES......

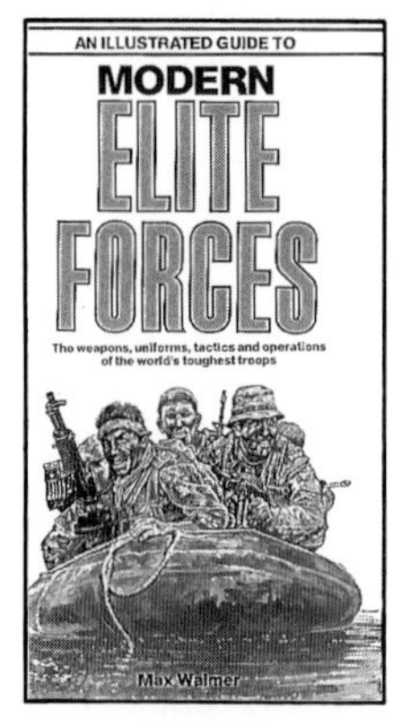

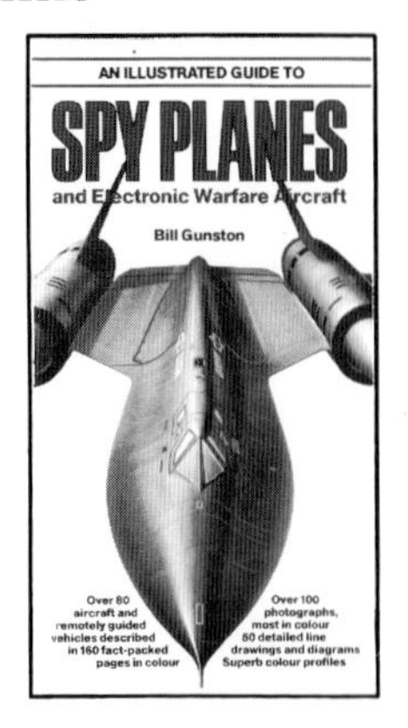

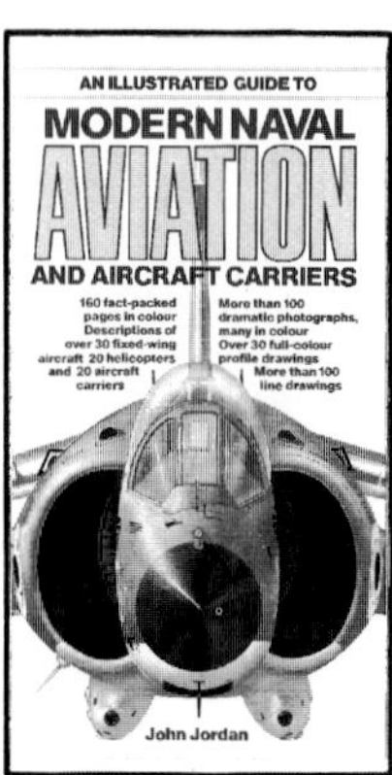

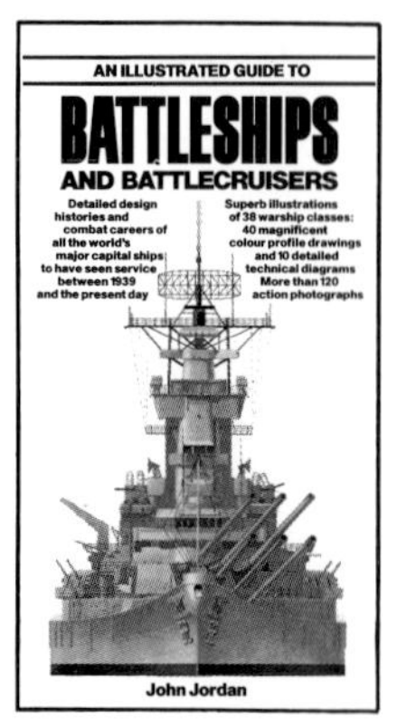

OTHER ILLUSTRATED MILITARY GUIDES NOW AVAILABLE.

Air War over Vietnam
Allied Fighters of World War II
Bombers of World War II
German, Italian and Japanese Fighters of World War II
Israeli Air Force
Military Helicopters
Modern Fighters and Attack Aircraft
Modern Soviet Air Force
Modern Soviet Ground Forces
Modern Soviet Navy
Modern Sub Hunters
Modern Submarines
Modern Tanks
Modern US Army
Modern US Navy
Modern Warships
Pistols and Revolvers
Rifles and Sub-Machine Guns
World War II Tanks

* Each has 160 fact-filled pages
* Each is colourfully illustrated with hundreds of action photographs and technical drawings
* Each contains concisely presented data and accurate descriptions of major international weapons
* Each represents tremendous value

If you would like further information on any of our titles please write to:

Publicity Dept. (Military Div.), Salamander Books Ltd.,
27 Old Gloucester Street, London WC1N 3AF

PRINTED IN BELGIUM BY proost INTERNATIONAL BOOK PRODUCTION